D1064808

COMPLEX STOCHASTIC PROCESSES
An Introduction to Theory and Application

COMPLEX STOCHASTIC PROCESSES

An Introduction to Theory and Application

Kenneth S. Miller
Riverside Research Institute, New York, New York

1974
Addison-Wesley Publishing Company, Inc.
Advanced Book Program
Reading, Massachusetts

London • Amsterdam • Don Mills, Ontario • Sydney • Tokyo

Library of Congress Cataloging in Publication Data

Miller, Kenneth S
Complex stochastic processes.

Bibliography: p.
1. Stochastic processes. 2. Functions of complex variables. 3. Estimation theory. I. Title.
QA274.M54 519.2 73-18094
ISBN 0-201-04744-6
ISBN 0-201-04745-4 (pbk.)

American Mathematical Society (MOS) Subject Classification Scheme (1970): 60-02, 60G, 60H, 62H, 62J, 60E05, 60G10, 60G15, 60G17, 60G35, 60H05, 60H10, 62E15, 62F10, 62H10, 62H15, 62J05, 62N15

Printed in the United States of America

CONTENTS

PREFACE

When the student of engineering or applied science is first exposed to stochastic processes, or noise theory, he is usually content to manipulate random variables formally as if they were ordinary functions. Sometime later the serious student becomes concerned about such problems as the validity of differentiating random variables and the interpretation of stochastic integrals, to say nothing of the usual worries associated with the interchange of the order of integration in multiple integrals. It is to this class of readers that this book is addressed. We attempt to analyze problems of the type just mentioned at an elementary yet rigorous level, and to adumbrate some of the physical applications of the theory.

Since complex processes are of interest in applied analysis (see, for example, [Ar], [CL], [D], [K], [KRR 1, 2], [MR 1, 2]), we have formulated our theory in terms of complex-valued random variables. In those cases where the extension from the real to the complex is not merely a notational change (for example, in the treatment of Gaussian processes) we carefully indicate where the differences lie. Chapter I is concerned with limiting operations involving random variables. We begin with a discussion of sequences of sets and probability spaces, and prove some basic theorems (for example, the Borel–Cantelli lemma). After defining random variables and related concepts, we embark on a detailed discussion and comparison of convergence with probability one, convergence in probability, and convergence in mean square.

The continuity and differentiability of random variables are analyzed in Chapter II. The appropriate setting for this discussion is found in the theory of stochastic processes. Accordingly, we begin with the elementary properties of stochastic processes, the covariance function, and the spectral distribution function. For the case of weakly stationary processes, sufficient conditions for almost everywhere continuity and mean square continuity may be simply expressed in terms of the covariance function, and by Bochner's theorem, in terms of the spectral distribution function. The simplifications that arise when the process is (complex) Gaussian are examined. Similar theorems are also established for differentiability almost everywhere and for differentiability in mean square.

The operation of the central limit theorem in physical phenomena naturally leads to Gaussian processes. We devote Chapter III to a discussion of complex Gaussian processes. The complex density function is introduced, and we obtain the generating function for a complex multidimensional Gaussian random variable. Finally, we use the complex density function to derive certain joint distributions and moments of amplitudes and phases of complex variates.

The remaining five chapters deal with various applications of complex stochastic processes. Chapter IV discusses stochastic integrals, including certain generalizations and infinite stochastic integrals. Linear differential equations with random forcing functions are solved in terms of the Green's function, and various relationships between the input and output covariance function and input and output spectral distribution function (when the process is wide-sense stationary) are also derived. The chapter concludes with a brief discussion of autoregressive processes.

The estimation of parameters in complex Gaussian stochastic processes forms the content of Chapter V. We consider both deterministic and random signals embedded in noise. The likelihood equations and Fisher information matrix are derived; and for some specific problems of broad interest the likelihood equations are solved and the Cramér–Rao lower bounds computed.

An interesting problem, with numerous physical applications, is the estimation of moments of the spectrum of a wide-sense stationary complex stochastic process. We sketch many of the numerous possible different approaches that could lead to a solution of this problem. In particular, the coherent, continuous, spectral approach to the theory of spectral moment estimation is analyzed with appropriate mathematical rigor in Chapter VI.

In Chapter VII certain problems in hypothesis testing are analyzed in detail. In particular, if $\boldsymbol{R}$ is a complex covariance matrix, we test the hypothesis $\boldsymbol{R} = \boldsymbol{R}_0$ versus the simple alternative $\boldsymbol{R} = \boldsymbol{R}_1$ in the pulse-pair case.

Finally, in Chapter VIII, certain elementary portions of the theory of linear least squares are cast in terms of complex quantities. The Gauss–Markov theorem is proved; and certain related results are treated. For example, we give necessary and sufficient conditions that the Markov estimate and least squares estimate be identical. Among other topics we consider are the efficiency of least squares estimates and their relationship to maximum likelihood estimates. We conclude with a discussion of regression curves. [Sections 1 and 3–7 of this chapter are taken mainly from an article by the author which appeared in SIAM Rev. **15,** 706-726, (1973), © 1973 by Society for Industrial and Applied Mathematics, and are reprinted with permission.]

Applications, of course, are a matter of taste, interest, or necessity for each worker in the field. We have found the foregoing material of particular use, and hope that it also will be of concern to others.

Prerequisites include an elementary course in the theory of functions of a real variable and familiarity with matrix notation. Naturally we also assume that the reader has an interest in, and some knowledge of, probability theory and random processes. An acquaintance with linear differential equations and Fourier transforms will prove useful, although not essential.

I take this opportunity to express my appreciation to many colleagues and friends who contributed to my interest, knowledge, and enthusiasm for complex stochastic processes. I am particularly indebted to Marvin M. Rochwarger, who suggested many of the applications and substantially contributed to their solution and physical interpretation.

KENNETH S. MILLER

COMPLEX STOCHASTIC PROCESSES
An Introduction to Theory and Application

I

CONVERGENCE OF RANDOM VARIABLES

0. Introduction

Arithmetic operations on finite ensembles of numbers or functions cause no mathematical complications. However, when we concern ourselves with infinite ensembles, new problems arise. Thus, in the treatment of the convergence of sequences, series, and integrals, supplementary definitions and techniques must be introduced. Naturally one would expect additional difficulties to arise when we attempt to analyze infinite ensembles of random variables. We propose to treat some of these problems in the present chapter.

We set the framework by discussing classes of sets, and in particular, sequences of sets. Various results associated with contracting and expanding sequences, and superior and inferior limits of sequences, are proved. The definition of a σ-algebra $\mathscr{F}$ of sets in a space Ω, and the definition of a functional P on $\mathscr{F}$, lead us to the definition of a probability space $\mathscr{P} = (\Omega, \mathscr{F}, P)$. Some results on the additivity of probability measure P (for example, the Borel–Cantelli lemma) are included. A random variable is defined, as usual, as a P-measurable function on $\mathscr{P}$. This immediately induces the consideration of the usual host clustered about this concept, for example, distribution function, mathematical expectation, and characteristic function.

In Sections 4, 5, and 6 we introduce the three fundamental modes of convergence: convergence with probability one, convergence in probability, and convergence in mean square. For convergence with probability one we mention some of the many different equivalent representations, including Cauchy sequences. For mean square convergence we treat the Hilbert space of mean zero random variables with finite variance, and the Cauchy and Loève criteria. Finally, in the last section we investigate some of the implications of these three modes of convergence, for example, convergence of subsequences. We also consider some counterexamples.

1. Sequences of Sets

Let Ω be an abstract space of points ω. If $A \subset \Omega$ is a set in Ω, then we denote the complement of A by $A^c = \Omega - A$. The empty set is $\emptyset$. Let $\mathscr{C}$ be a class of sets in Ω. Then

$$\bigcup_{A\in\mathscr{C}} A = \{\omega|\omega \in A \quad \text{for at least one} \quad A \in \mathscr{C}\}$$

and

$$\bigcap_{A\in\mathscr{C}} A = \{\omega|\omega \in A \quad \text{for all} \quad A \in \mathscr{C}\} \quad .$$

Thus it is easily seen that

$$\left(\bigcup_{A\in\mathscr{C}} A\right)^c = \bigcap_{A\in\mathscr{C}} A^c \quad , \tag{1.1}$$

and

$$\left(\bigcap_{A\in\mathscr{C}} A\right)^c = \bigcup_{A\in\mathscr{C}} A^c \quad . \tag{1.2}$$

Now let $\{A_n\}$, where $A_n \subset \Omega$ for all n, be a *sequence* of sets in Ω. Then we assert that

$$\bigcap_{m=1}^{\infty} \bigcup_{k=m}^{\infty} A_k = \left(\bigcup_{m=1}^{\infty} \bigcap_{k=m}^{\infty} A_k^c\right)^c \quad . \tag{1.3}$$

For, let $B_m = \bigcup_{k=m}^{\infty} A_k$. Then from (1.2)

$$\bigcap_{m=1}^{\infty} B_m = \left(\bigcup_{m=1}^{\infty} B_m^c\right)^c \tag{1.4}$$

and from (1.1)

$$B_m^c = \left(\bigcup_{k=m}^{\infty} A_k\right)^c = \bigcap_{k=m}^{\infty} A_k^c \quad . \tag{1.5}$$

Thus (1.4) and (1.5) imply (1.3).

We define (1.3) as the *limit superior* of the sequence $\{A_n\}$ and write

$$A^* = \limsup_k A_k = \bigcap_{m=1}^{\infty} \bigcup_{k=m}^{\infty} A_k \quad . \tag{1.6}$$

Similarly, the *limit inferior* of the sequence $\{A_n\}$ is defined as

$$A_* = \liminf_k A_k = \bigcup_{m=1}^{\infty} \bigcap_{k=m}^{\infty} A_k \quad . \tag{1.7}$$

Now

$$B_m = \bigcup_{k=m}^{\infty} A_k = \{\omega | \omega \in A_k \quad \text{for at least one} \quad k \geqq m\}$$

and hence

$$A^* = \bigcap_{m=1}^{\infty} B_m = \{\omega | \omega \in A_k \quad \text{for infinitely many} \quad k\} \quad . \tag{1.8}$$

Equation (1.8) is a convenient interpretation of the limit superior in probabilistic arguments.

If

$$\Omega \supset C_k \supset C_{k+1} \quad , \qquad k = 1, 2, \cdots \quad ,$$

then we call $\{C_n\}$ a *contracting sequence*. For a contracting sequence

$$\bigcup_{k=m}^{\infty} C_k = C_m$$

and

$$\limsup_k C_k = \bigcap_{m=1}^{\infty} \bigcup_{k=m}^{\infty} C_k = \bigcap_{m=1}^{\infty} C_m \quad .$$

Let $D = \bigcap_{k=m}^{\infty} C_k$. Then D is independent of m, and we may write

$$\bigcup_{m=1}^{\infty} D = D \quad .$$

Thus

$$\liminf_k C_k = \bigcup_{m=1}^{\infty} \bigcap_{k=m}^{\infty} C_k = \bigcup_{m=1}^{\infty} D = D = \bigcap_{k=1}^{\infty} C_k$$

and

$$\limsup_k C_k = \liminf_k C_k \quad .$$

We write this common value as

$$\lim_k C_k = \bigcap_{k=1}^{\infty} C_k \quad .$$

Similarly, an *expanding sequence* $\{E_n\}$ is defined by the property

$$E_k \subset E_{k+1} \subset \Omega, \qquad k = 1, 2, \cdots \quad ,$$

and we may show that

$$\limsup_k E_k = \liminf_k E_k = \bigcup_{k=1}^{\infty} E_k \quad .$$

We call this common value $\lim_k E_k$:

$$\lim_k E_k = \bigcup_{k=1}^{\infty} E_k \quad .$$

Formally stated:

Theorem 1.1. If $\{C_n\}$ is a contracting sequence in a space Ω, then

$$\lim_n C_n = \bigcap_{n=1}^{\infty} C_n \tag{1.9}$$

and if $\{E_n\}$ is an expanding sequence,

$$\lim_n E_n = \bigcup_{n=1}^{\infty} E_n \quad . \tag{1.10}$$

If the members of a sequence $\{B_n\}$ in Ω are all identical, say $B_n = B_0$ for all n, then by definition of limits

$$\limsup_n B_n = \liminf_n B_n = \lim_n B_n = B_0 \quad .$$

Lemma 1.1. Let $\{C_n\}$ be a contracting sequence in a space Ω. Let $C_n \subset B_0 \subset \Omega$ for all n. Then

$$\lim_n (B_0 - C_n) = B_0 - \lim_n C_n \quad .$$

Proof. Since $\{C_n\}$ is a contracting sequence, $\{B_0 - C_n\}$ is an expanding sequence. Thus by (1.10),

$$\lim_n (B_0 - C_n) = \bigcup_{n=1}^{\infty} (B_0 - C_n) = \bigcup_{n=1}^{\infty} (B_0^c \cup C_n)^c$$

and by (1.2)

$$\lim_n (B_0 - C_n) = \left(\bigcap_{n=1}^{\infty} (B_0^c \cup C_n)\right)^c \quad .$$

Now it is easily verified that

$$\bigcap_{n=1}^{\infty} (B_0^c \cup C_n) = B_0^c \cup \left(\bigcap_{n=1}^{\infty} C_n\right) \quad ,$$

and hence

$$\lim_n (B_0 - C_n) = B_0 - \bigcap_{n=1}^{\infty} C_n = B_0 - \lim_n C_n$$

by (1.9).

Let $\{A_n\}$ be an arbitrary sequence of sets in Ω. Then

$$\limsup_k A_k = \bigcap_{m=1}^{\infty} \bigcup_{k=m}^{\infty} A_k \quad .$$

Let $C_m = \bigcup_{k=m}^{\infty} A_k$. Then $\{C_m\}$ is a contracting sequence and by (1.9)

$$\limsup_k A_k = \bigcap_{m=1}^{\infty} C_m = \lim_m C_m = \lim_m \bigcup_{k=m}^{\infty} A_k \quad .$$

Similarly .

$$\liminf_k A_k = \bigcup_{m=1}^{\infty} \bigcap_{k=m}^{\infty} A_k = \bigcup_{m=1}^{\infty} E_m$$

where $E_m = \bigcap_{k=m}^{\infty} A_k$ and $\{E_m\}$ is an expanding sequence. Thus by (1.10),

$$\liminf_k A_k = \lim_m E_m = \lim_m \bigcap_{k=m}^{\infty} A_k \quad .$$

Now in the notation of the previous paragraph,

$$\left(\bigcap_{m=1}^{\infty} C_m\right)^c = \bigcup_{m=1}^{8} C_m^c$$

by (1.2). But $\{C_m\}$ is a contracting sequence and $\{C_m^c\}$ is an expanding sequence. Thus by Theorem 1.1

$$\left(\lim_m C_m\right)^c = \lim_m C_m^c \tag{1.11}$$

This result then implies:

Theorem 1.2. Let $\{A_n\}$ be a sequence of sets in a space Ω. Then

$$\left(\lim_m \bigcup_{k=m}^{\infty} A_k\right)^c = \lim_m \bigcap_{k=m}^{\infty} A_k^c \quad .$$

We have used the fact that

$$C_m^c = \left(\bigcup_{k=m}^{\infty} A_k\right)^c = \bigcap_{k=m}^{\infty} A_k^c$$

[see (1.1)].

2. Probability Space

Let $\mathscr{F}$ be a class of sets, in an abstract space Ω, with the following properties:

(i) $\emptyset \in \mathscr{F}$

(ii) If $A \in \mathscr{F}$, then $A^c \in \mathscr{F}$

(iii) If $\{A_n\}$ is a sequence of sets in $\mathscr{F}$, then $\bigcup_{n=1}^{\infty} A_n \in \mathscr{F}$.

Then we say $\mathscr{F}$ is a *sigma-algebra* or a *Borel field.*

Since $A - B = (A^c \cup B)^c$, we conclude that if A and B are in $\mathscr{F}$, then so is $A - B$. Now let $\{A_n\}$ be a sequence of sets in $\mathscr{F}$, and let $A = \bigcup_{n=1}^{\infty} A_n$. Then the identity

$$\bigcap_{n=1}^{\infty} A_n = A - \bigcup_{n=1}^{\infty} (A - A_n) \tag{2.1}$$

implies that

$$\bigcap_{n=1}^{\infty} A_n \in \mathscr{F} \quad .$$

Finally, since $\emptyset^c = \Omega$ and $\emptyset \in \mathscr{F}$, we infer that $\Omega \in \mathscr{F}$.

Let P be a mapping of $\mathscr{F}$ into the unit interval [0,1] with the properties that

(i) $P[A] \geqq 0$ for all $A \in \mathscr{F}$

(ii) $P[\Omega] = 1$

(iii) If $\{A_n\}$ is a sequence of sets in $\mathscr{F}$, then

$$P[\bigcup_{n=1}^{\infty} A_n] = \sum_{n=1}^{\infty} P[A_n]$$

provided $A_j \cap A_k = \emptyset$ for $j \neq k$.

Condition (iii) implies that P is a *completely additive set function.* Conditions (i) and (iii) imply that P is a *measure function.* Conditions (i), (ii), and (iii) imply that P is a *probability measure.*

Now let Ω be an abstract space, $\mathscr{F}$ a class of subsets of Ω, and P a functional defined on $\mathscr{F}$. More specifically, let Ω be a collection of abstract points ω, let $\mathscr{F}$ be a sigma-algebra on Ω, and let P be a probability measure on $\mathscr{F}$. Then we call the triplet $\mathscr{P} = (\Omega, \mathscr{F}, P)$ a *probability space.*

Lemma 2.1. If A and B are sets on a probability space $\mathscr{P}$, and $B \subset A$, then

$$P[A - B] = P[A] - P[B] \quad . \tag{2.2}$$

Proof. Since

$$(A - B)^c = A^c \cup B$$

and $A^c \cap B = \emptyset$, we have

$$1 - P[A - B] = P[(A - B)^c] = P[A^c \cup B] = P[A^c] + P[B]$$
$$= 1 - P[A] + P[B]$$

or

$$P[A - B] = P[A] - P[B] \quad .$$

Corollary 2.1. If $B \subset A$, then $P[B] \leqq P[A]$.

Proof. Since $P[A - B]$ is nonnegative, (2.2) implies

$$P[A] - P[B] \geqq 0 \quad .$$

Theorem 2.1. Let $\{E_n\}$ be an expanding sequence of sets defined on a probability space $\mathscr{P}$. Then

$$P[\lim_n E_n] = \lim_{n\to\infty} P[E_n] \quad . \tag{2.3}$$

Proof. Let

$$F_1 = E_1 \quad ,$$

$$F_k = E_k - E_{k-1} \quad , \qquad k = 2, 3, \cdots \quad .$$

Then $\{F_k\}$ is a sequence of disjoint sets. By (iii) of the definition of a probability measure,

$$P[\bigcup_{k=1}^{\infty} F_k] = \sum_{k=1}^{\infty} P[F_k] \quad . \tag{2.4}$$

Now it is easily verified that

$$\bigcup_{k=1}^{\infty} F_k = \bigcup_{k=1}^{\infty} E_k \quad .$$

Thus

$$P[\bigcup_{k=1}^{\infty} F_k] = P[\bigcup_{k=1}^{\infty} E_k] = P[\lim_k E_k] \tag{2.5}$$

by Theorem 1.1.

Also

$$P[F_1] = P[E_1]$$

and

$$P[F_k] = P[E_k] - P[E_{k-1}] \quad , \qquad k = 2, 3, \cdots$$

by Lemma 2.1. Therefore

$$\sum_{k=1}^{\infty} P[F_k] = \lim_{n\to\infty} \sum_{k=1}^{n} P[F_k] = \lim_{n\to\infty} P[E_n] \quad . \tag{2.6}$$

Substituting (2.5) and (2.6) into (2.4) yields (2.3).

Theorem 2.2. Let $\{C_n\}$ be a contracting sequence of sets defined on a probability space $\mathscr{P}$. Then

$$P[\lim_n C_n] = \lim_{n\to\infty} P[C_n] \quad . \tag{2.7}$$

Proof. Since $\{C_n\}$ is a contracting sequence, $C_n \subset C_1$ for all n, and by Theorem 1.1

$$\lim_n C_n = \bigcap_{n=1}^{\infty} C_n \subset C_1 \quad .$$

Now by Lemmas 1.1 and 2.1

$$\begin{aligned} P[\lim_n (C_1 - C_n)] &= P[C_1 - \lim_n C_n] \\ &= P[C_1] - P[\lim_n C_n] \quad . \end{aligned} \tag{2.8}$$

But since $\{C_1 - C_n\}$ is an expanding sequence,

$$P[\lim_n (C_1 - C_n)] = \lim_{n\to\infty} P[C_1 - C_n] = P[C_1] - \lim_{n\to\infty} P[C_n] \tag{2.9}$$

by Theorem 2.1 and Lemma 2.1. Comparing (2.8) and (2.9) yields (2.7).

Corollary 2.2. Let $\{A_n\}$ be a sequence of sets on a probability space $\mathscr{P}$. Then

$$P[A_*] = P[\lim_m \bigcap_{k=m}^{\infty} A_k] = \lim_{m\to\infty} P[\bigcap_{k=m}^{\infty} A_k] \quad . \tag{2.10}$$

Proof. By definition the limit inferior is

$$A_* = \bigcup_{m=1}^{\infty} \bigcap_{k=m}^{\infty} A_k \quad .$$

Now let

$$E_m = \bigcap_{k=m}^{\infty} A_k \quad .$$

Then $\{E_m\}$ is an expanding sequence. By Theorem 2.1

$$P[\lim_m E_m] = \lim_{m\to\infty} P[E_m] \tag{2.11}$$

while Theorem 1.1 yields

$$P[\lim_m E_m] = P[\bigcup_{m=1}^{\infty} E_m] \quad . \tag{2.12}$$

These equations imply (2.10).

Corollary 2.3. Let $\{A_n\}$ be a sequence of sets on a probability space $\mathscr{P}$. Then

$$P[A^*] = P[\lim_m \bigcup_{k=m}^{\infty} A_k] = \lim_{m\to\infty} P[\bigcup_{k=m}^{\infty} A_k] \quad .$$

Proof. By definition the limit superior is

$$A^* = \bigcap_{m=1}^{\infty} \bigcup_{k=m}^{\infty} A_k \quad .$$

Now let

$$C_m = \bigcup_{k=m}^{\infty} A_k \quad .$$

Then $\{C_m\}$ is a contracting sequence. Theorems 2.2 and 1.1 complete the proof.

Theorem 2.3. Let $\{A_n\}$ be a sequence of sets defined on a probability space $\mathscr{P}$. Then

$$P[\bigcup_{n=1}^{\infty} A_n] \leqq \sum_{n=1}^{\infty} P[A_n] \quad .$$

Proof. Let

$$B_1 = A_1$$

$$B_k = A_k - \bigcup_{j=1}^{k-1} A_j \quad , \qquad k = 2, 3, \cdots \quad .$$

Then $B_j \cap B_k = \emptyset$ for $j \neq k$ and

$$\bigcup_{n=1}^{\infty} A_n = \bigcup_{n=1}^{\infty} B_n \quad .$$

By definition of a probability measure

$$P[\bigcup_{n=1}^{\infty} A_n] = P[\bigcup_{n=1}^{\infty} B_n] = \sum_{n=1}^{\infty} P[B_n] \quad .$$

But since $B_k \subset A_k$ for all k, Corollary 2.1 implies $P[B_k] \leqq P[A_k]$. Thus

$$\sum_{n=1}^{\infty} P[B_n] \leqq \sum_{n=1}^{\infty} P[A_n] \quad .$$

Theorem 2.4 (Borel–Cantelli Lemma). Let $\{A_n\}$ be a sequence of sets defined on a probability space $\mathscr{P}$. Let A^* be the limit superior of the sequence $\{A_n\}$. Then if

$$\sum_{n=1}^{\infty} P[A_n] < \infty \quad ,$$

we have

$$P[A^*] = 0 \quad .$$

Proof. By Corollary 2.3 and Theorem 2.3

$$P[A^*] = \lim_{m\to\infty} P[\bigcup_{k=m}^{\infty} A_k] \leqq \lim_{m\to\infty} \sum_{k=m}^{\infty} P[A_k] \quad .$$

But $\sum_{k=1}^{\infty} P[A_k]$ converges by hypothesis. Thus $P[A^*] = 0$.

3. Random Variables

Let $\mathscr{P} = (\Omega, \mathscr{F}, P)$ be a probability space. Let X be a mapping of Ω into the real numbers such that

$$S_x = \{\omega \mid X(\omega) \leqq x\} \in \mathscr{F} \tag{3.1}$$

for all real x. Then we say X is a (real) *random variable* defined on $\mathscr{P}$. Thus a random variable is simply a P-measurable function on $\mathscr{P}$. We call

$$F(x) = P[S_x] = P[X \leqq x] \tag{3.2}$$

the *distribution function* of X.

Since $S_x \subset S_{x'}$ for $x < x'$, Corollary 2.1 implies that $P[S_x] \leqq P[S_{x'}]$ or

$$F(x) \leqq F(x') \quad , \qquad x < x' \quad .$$

Hence F is a monotonically increasing function on $(-\infty, \infty)$. Now for *any* monotone increasing function G, both $G(\xi+)$ and $G(\xi-)$ exist for all ξ in its domain, $G(\xi-) \leqq G(\xi) \leqq G(\xi+)$, and the set of points at which G is discontinuous is at most enumerable. Since we have defined $F(x)$ as $P[X \leqq x]$, we see that $F(x) = F(x+)$, that is, F is right continuous. (If we had defined F as $F(x) = P[X < x]$, then F would be left continuous.)

Let $\{x_n\}$ be a decreasing sequence of numbers converging to $-\infty$. Then $\{S_{x_n}\}$ is a contracting sequence of sets and

$$F(-\infty) = \lim_{x_n \to -\infty} F(x_n) = \lim_{n\to\infty} P[S_{x_n}] = P[\lim_n S_{x_n}]$$

by Theorem 2.2. But by Theorem 1.1

$$F(-\infty) = P[\bigcap_{n=1}^{\infty} S_{x_n}] \quad .$$

Since $\bigcap_{n=1}^{\infty} S_{x_n} = \emptyset$, we have

$$F(-\infty) = P[\emptyset] = 0 \quad .$$

Similarly, if $\{x'_n\}$ is an increasing sequence of numbers coverging to $+\infty$, then $\{S_{x'_n}\}$ is an expanding sequence of sets. Theorems 2.1 and 1.1 then imply

$$F(\infty) = \lim_{n\to\infty} P[S_{x'_n}] = P[\lim_n S_{x'_n}] = P[\bigcup_{n=1}^{\infty} S_{x'_n}] = P[\Omega] = 1 \quad .$$

Thus a distribution function F has the following properties

(i) F is monotone increasing

(ii) $0 \leqq F(x) \leqq 1$

(iii) $F(-\infty) = 0 \quad , \quad F(\infty) = 1$

(iv) F is right continuous

(v) the set of points on which F is discontinuous is at most countable.

Now let g be a function defined on $(-\infty, \infty)$ and let F be a distribution function of a random variable X. Then if

$$\lim_{\substack{a\to-\infty \\ b\to\infty}} \int_a^b g(x)dF(x)$$

exists, we write this limit as

$$\mathscr{E}g(X) = \int_{-\infty}^{\infty} g(x)dF(x) \tag{3.3}$$

and call the improper Riemann–Stieltjes integral of (3.3) the *mathematical expectation* of g.

Lemma 3.1. Let g be (Riemann) integrable and bounded on $(-\infty, \infty)$. Let F be a distribution function. Then $\mathscr{E}g$ exists.

Proof. Let $\varepsilon > 0$ be assigned and choose an $A > 0$ such that

$$\int_a^{a'} dF(x) < \tfrac{1}{2}\varepsilon \qquad \text{and} \qquad \int_b^{b'} dF(x) < \tfrac{1}{2}\varepsilon$$

for all $a < a' < -A$ and $b' > b > A$. Then

$$\left| \int_a^{b'} g(x)dF(x) - \int_{a'}^{b} g(x)dF(x) \right| < M\varepsilon$$

where $M \geqq |g(x)|$ on $(-\infty, \infty)$. By the Cauchy convergence theorem for real numbers $\mathscr{E}g$ exists.

If F is the distribution function of the random variable X, then

$$\mathscr{E}e^{itx} = \phi(t) = \int_{-\infty}^{\infty} e^{itx}\, dF(x) \tag{3.4}$$

is called the *characteristic function* of X. By Lemma 3.1 we see that ϕ always exists. Furthermore we conclude that:

Theorem 3.1. The characteristic function ϕ of a random variable X is uniformly continuous on $(-\infty, \infty)$.

Proof. By (3.4)

$$\tfrac{1}{2}|\phi(t+h) - \phi(t)| \leqq \int_{-\infty}^{\infty} |\sin \tfrac{1}{2} hx|\, dF(x)$$

where F is the distribution function of X. For any $A > 0$

$$\tfrac{1}{2}\, |\phi(t+h) - \phi(t)| \leqq \int_{-\infty}^{-A} dF(x) + \int_{-A}^{A} |\sin \tfrac{1}{2} hx|\, dF(x) + \int_{A}^{\infty} dF(x) \quad .$$

Now we can choose an $A > 0$ so large and then choose an $h \neq 0$ so small (in absolute value) that the above three integrals are less than any preassigned positive number. Thus $\phi(t)$ is uniformly continuous for all t.

A function K defined on a subset T of the real numbers is said to be *non-negative definite* if $K(t) = \bar{K}(-t)$ for all $t \in T$ and if

$$\sum_{j,k=1}^{n} K(t_j - t_k)\lambda_j\bar{\lambda}_k \geqq 0$$

for all sets of distinct points t_j, $1 \leqq j \leqq n$, in T; all complex numbers λ_j, $1 \leqq j \leqq n$; and all positive integers n.

Theorem 3.2. The characteristic function ϕ of a random variable X is nonnegative definite on $(-\infty, \infty)$.

Proof. The theorem follows immediately from the identity

$$\sum_{j,k=1}^{n} \phi(t_j - t_k)\, \lambda_j \bar{\lambda}_k = \int_{-\infty}^{\infty} \left| \sum_{j=1}^{n} [\exp(it_j x)] \lambda_j \right|^2 dF(x) \geqq 0$$

where F is the distribution function of X.

We also state for future reference [Cr, page 96]:

Theorem 3.3. Let $\{F_n\}$ be a sequence of distribution functions with corresponding characteristic functions $\{\phi_n\}$. Then a necessary and sufficient condition that $\{F_n\}$ converge to a distribution function F is that $\{\phi_n(t)\}$ converge for all t to a function $\phi(t)$ continuous at $t = 0$. Then the function ϕ is the characteristic function corresponding to F.

Let X be a random variable defined on $\mathscr{P} = (\Omega, \mathscr{F}, P)$ and let Y be another random variable defined on the same probability space. Then by definition of random variable,

$$T_y = \{\omega \mid Y(\omega) \leqq y\} \in \mathscr{F} \quad .$$

Since $\mathscr{F}$ is a σ-algebra, $S_x \cap T_y \in \mathscr{F}$ [see (3.1)]. We call

$$\begin{aligned} F(x, y) = P[S_x \cap T_y] &= P[\{\omega \mid X(\omega) \leqq x \quad \text{and} \quad Y(\omega) \leqq y\}] \\ &= P[X \leqq x,\ Y \leqq y] \end{aligned}$$

the *joint distribution function* of X and Y. For any function $g(\cdot,\cdot)$ we define the expectation of g as

$$\mathscr{E} g(X,Y) = \int_{-\infty}^{\infty} \int_{-\infty}^{\infty} g(x, y)\, dF(x, y) \quad ,$$

provided the integral exists.

Let F be the distribution function of the random variable X defined on $\mathscr{P}$. If the derivative F' of F exists, we call F' the *frequency function* or *probability density function* of X. If $F(x, y)$ is the joint distribution function of X and Y, we call

$$f(x, y) = \frac{\partial^2}{\partial x \partial y} F(x, y)$$

(if it exists) the *joint frequency function* or *joint probability density function* of X and Y. The extension of the definition of distribution function and frequency function to any finite number of random variables is immediate.

If X and Y are two random variables (defined on the same probability space) which are equal with probability one, that is, if

$$P[\{\omega | X(\omega) = Y(\omega)\}] = 1 \quad , \tag{3.5}$$

then we say X and Y are *equivalent*, Thus, equivalent random variables differ (at most) on an ω-set of P-measure zero.

In subsequent work we shall consider *complex* random variables. Thus if X and Y are as defined above, $\xi = X + iY$ is a complex random variable. Clearly, the properties of ξ may be deduced from those of X and Y.

4. Convergence with Probability One

Let $\{\xi_n\}$ be a sequence of real or complex random variables (P-measurable functions) defined on a probability space $\mathscr{P} = (\Omega, \mathscr{F}, P)$. Let ξ be a random variable defined on $\mathscr{P}$. If for every $\varepsilon > 0$, $\delta > 0$ there exists an integer N such that

$$P[\{\omega | |\xi_n(\omega) - \xi(\omega)| > \varepsilon \quad \text{for at least one} \quad n \geqq N\}] < \delta \quad , \tag{4.1}$$

then we say $\{\xi_n\}$ *converges with probability one to* ξ. Alternate terms are *convergence almost everywhere, almost certain convergence, strong convergence, almost sure convergence*. Thus in (4.1) we are considering the probability measure of the event $\{\omega | |\xi_n(\omega) - \xi(\omega)| > \varepsilon$ for at least one $n \geqq N\}$. For simplicity in notation [just as we did in (3.2)] we drop the braces and the ω to write (4.1) as

$$P[|\xi_n - \xi| > \varepsilon \quad \text{for at least one} \quad n \geqq N] < \delta \quad . \tag{4.2}$$

By definition of limit we may write (4.1) or (4.2) as

$$\lim_{N\to\infty} P[|\xi_n - \xi| > \varepsilon \quad \text{for at least one} \quad n \geqq N] = 0 \quad . \tag{4.3}$$

Since

$$\{\omega | |\xi_n(\omega) - \xi(\omega)| > \varepsilon \quad \text{for at least one} \quad n \geqq N\} = \bigcup_{n=N}^{\infty} \{\omega | |\xi_n(\omega) - \xi(\omega)| > \varepsilon\}$$

another equivalent form is

$$\lim_{N\to\infty} P[\bigcup_{n=N}^{\infty} \{|\xi_n - \xi| > \varepsilon\}] = 0 \quad . \tag{4.4}$$

(See also Corollary 2.3.) By (1.1)

$$\left(\bigcup_{n=N}^{\infty} \{|\xi_n - \xi| > \varepsilon\}\right)^c = \bigcap_{n=N}^{\infty} \{|\xi_n - \xi| \leqq \varepsilon\}$$

and we have the further alternative form

$$\lim_{N\to\infty} P[\bigcap_{n=N}^{\infty} \{|\xi_n - \xi| \leqq \varepsilon\}] = 1 \quad . \tag{4.5}$$

Continuing, the relationship

$$\bigcap_{n=N}^{\infty} \{\omega \mid |\xi_n(\omega) - \xi(\omega)| \leqq \varepsilon\}$$

$$= \{\omega \mid |\xi_n(\omega) - \xi(\omega)| \leqq \varepsilon \quad \text{for all} \quad n \geqq N\}$$

implies

$$\lim_{N\to\infty} P[|\xi_n - \xi| \leqq \varepsilon \quad \text{for all} \quad n \geqq N] = 1 \quad . \tag{4.6}$$

Thus (4.1)–(4.6) are all equivalent expressions of the fact that $\{\xi_n\}$ converges with probability one to ξ.

We may go even further. Equation (4.6) is equivalent to: For every $\varepsilon > 0$ and for almost all $\omega \in \Omega$ there exists an $N \equiv N(\omega, \varepsilon)$ such that

$$|\xi_n(\omega) - \xi(\omega)| \leqq \varepsilon \tag{4.7}$$

for all $n \geqq N$. In terms of "limit" terminology we may write

$$\lim_{n\to\infty} \xi_n = \xi \qquad [\text{a.e.}] \tag{4.8}$$

We prefer to drop the "a.e." and write

$$\lim_{n\to\infty} \xi_n = \xi \tag{4.9}$$

if indeed $\{\xi_n\}$ converges with probability one to ξ. Thus another valid description of convergence with probability one is

$$P[\{\omega \mid \lim_{n\to\infty} \xi_n - \xi = 0\}] = 1 \quad . \tag{4.10}$$

If we recall [Mu, pages 154, 222]

Theorem 4.1. If $\{\xi_n\}$ is a Cauchy sequence of random variables defined on a probability space $\mathscr{P}$, and if $\{\xi_n\}$ converges with probability one to ξ, then ξ is a P-measurable function,

then we see that if $\{\xi_n\}$ is a Cauchy sequence which converges with probability one, that is, if for every positive ε and δ there is an N such that

$$P[|\xi_n - \xi_m| > \varepsilon \quad \text{for at least one} \quad m \geqq n] < \delta \tag{4.11}$$

for all $n \geqq N$, then there exists a random variable ξ defined on $\mathscr{P}$ such that $\{\xi_n\}$ converges with probability one to ξ, Furthermore, the hypothesis that

ξ be a P-measurable function used in formulating (4.1)–(4.10) may be dropped.

In general, by Corollary 2.1 and Theorem 2.3,

$$P[|\xi_N - \xi| > \varepsilon] \leqq P[\bigcup_{n=N}^{\infty} \{|\xi_n - \xi| > \varepsilon\}] \leqq \sum_{n=N}^{\infty} P[|\xi_n - \xi| > \varepsilon] \quad . \tag{4.12}$$

Thus [see (4.4)] a sufficient condition that $\{\xi_n\}$ converge with probability one to ξ is that the series

$$\sum_{n=1}^{\infty} P[|\xi_n - \xi| > \varepsilon] \tag{4.13}$$

converge for every $\varepsilon > 0$. By the Tschebyscheff inequality

$$P[|\xi_n - \xi| > \varepsilon] \leqq (1/\varepsilon^2)\, \mathscr{E} |\xi_n - \xi|^2 \quad . \tag{4.14}$$

Thus a weaker sufficient condition for the convergence of $\{\xi_n\}$ to ξ with probability one is that

$$\sum_{n=1}^{\infty} \mathscr{E} |\xi_n - \xi|^2 < \infty \quad . \tag{4.15}$$

A useful result in later work is

Lemma 4.1. Let $\{\xi_n\}$ be a sequence of random variables defined on a probability space $\mathscr{P}$. Let $\{\delta_n\}$ and $\{\varepsilon_n\}$ be sequences of positive numbers such that $\sum_{n=1}^{\infty} \delta_n < \infty$ and $\sum_{n=1}^{\infty} \varepsilon_n < \infty$. Then if

$$P[|\xi_{n+1} - \xi_n| > \varepsilon_n] < \delta_n \quad , \qquad n = 1, 2, \cdots$$

there exists a P-measurable function ξ such that $\{\xi_n\}$ converges with probability one to ξ.

Proof. Let A^* be the limit superior of the sequence of events $\{\omega \,|\, |\xi_{k+1}(\omega) - \xi_k(\omega)| > \varepsilon_k\}$. Then [see (1.8)]

$$P[A^*] = P[|\xi_{k+1} - \xi_k| > \varepsilon_k \quad \text{for infinitely many} \quad k] \quad .$$

But by hypothesis

$$\sum_{k=1}^{\infty} P[|\xi_{k+1} - \xi_k| > \varepsilon_k] < \sum_{k=1}^{\infty} \delta_k < \infty \quad .$$

Thus by the Borel–Cantelli lemma (Theorem 2.4), $P[A^*] = 0$.

Thus, except on an ω-set of P-measure zero we can find an $N \equiv N(\omega)$ such that

$$|\xi_{n+1} - \xi_n| \leqq \varepsilon_n$$

for all $n > N$. Otherwise there would be an infinite number of integers k such that $|\xi_{k+1} - \xi_k| > \varepsilon_k$. But we have just shown that the probability of this event is zero. Thus the sequence $\{\xi_n\}$ converges, say to ξ, a.e. (Theorem 4.1), and ξ is P-measurable.

5. Convergence in Probability

Let $\{\xi_n\}$ be a sequence of real or complex random variables (P-measurable functions) defined on a probability space $\mathscr{P} = (\Omega, \mathscr{F}, P)$. Let ξ be a random variable defined on $\mathscr{P}$. If for every $\varepsilon > 0, \delta > 0$ there exists an integer N such that

$$P[\{\omega \mid |\xi_n(\omega) - \xi(\omega)| > \varepsilon\}] < \delta \tag{5.1}$$

for all $n > N$, then we say $\{\xi_n\}$ *converges in probability to* ξ. Alternate terms are *stochastic convergence*, *convergence in P-measure*, *convergence in measure*. For simplicity we often write (5.1) as

$$P[|\xi_n - \xi| > \varepsilon] < \delta \tag{5.2}$$

for all $n > N$. By definition of limit we may write (5.1) or (5.2) as

$$\lim_{n\to\infty} P[|\xi_n - \xi| > \varepsilon] = 0 \quad . \tag{5.3}$$

In terms of "limit" terminology,

$$\lim_{n\to\infty} \xi_n = \xi \qquad [\text{meas}] \quad . \tag{5.4}$$

We prefer to drop the "meas" and use probabilistic language to write

$$\operatorname*{p\,lim}_{n\to\infty} \xi_n = \xi \tag{5.5}$$

if indeed $\{\xi_n\}$ converges in probability to ξ. Equations (5.1)–(5.5) are all equivalent expressions of the fact that $\{\xi_n\}$ converges in probability to ξ.

If we recall [Mu, page 226]

Theorem 5.1. Let $\{\xi_n\}$ be a sequence of P-measurable functions defined on a probability space $\mathscr{P}$. If a Cauchy sequence converges in probability, then there exists a P-measurable function ξ such that $\{\xi_n\}$ converges in probability to ξ,

then we see that if $\{\xi_n\}$ is a Cauchy sequence which converges in probability, that is, if for every positive ε and δ there is an N such that

$$P[|\xi_n - \xi_m| > \varepsilon] < \delta$$

for all n, $m \geqq N$, then there exists a random variable ξ defined on $\mathscr{P}$ such that $\{\xi_n\}$ converges in probability to ξ. Furthermore, the hypothesis that ξ be a P-measurable function used in formulating (5.1)–(5.5) may be dropped.

Suppose $\{X_n\}$ is a sequence of real random variables defined on $\mathscr{P}$. Let F_n be the distribution function of X_n. If the sequence $\{F_n\}$ converges to a distribution function F at every point of continuity of F, then we say $\{X_n\}$ *converges in distribution.* Convergence in distribution does not imply that the sequence of random variables $\{X_n\}$ converges in any reasonable sense, say with probability one or in probability. For example, suppose that the $\{X_n\}$ are arbitrary random variables in $\mathscr{P}$, each with the same distribution function. However, if $\{X_n\}$ converges in probability to X *and* $\{X_n\}$ converges in distribution, then the distribution function of X is the limit of the sequence of distribution functions $\{F_n\}$ of $\{X_n\}$. Formally stated:

Theorem 5.2. Let $\{X_n\}$ be a sequence of random variables, defined on a probability space $\mathscr{P}$, which converges in probability to X. Let F_n be the distribution function of X_n and let $\{F_n\}$ converge to F. Let G be the distribution function of X. Then at every point x_0 where F is continuous,

$$F(x_0) = G(x_0) \quad .$$

Proof. For any real number x and any $\varepsilon > 0$ define the sets:

$$A_n = \{\omega \mid X_n(\omega) \leqq x - \varepsilon\}$$

$$B_n = \{\omega \mid X_n(\omega) \leqq x + \varepsilon\}$$

$$C = \{\omega \mid X(\omega) \leqq x\}$$

$$D_n = \{\omega \mid |X_n(\omega) - X(\omega)| \leqq \varepsilon\} \quad .$$

Then it is easily verified that

$$A_n \cap D_n \subset C \cap D_n \subset B_n \cap D_n \quad .$$

Now by Corollary 2.1

$$P[A_n] - P[A_n \cup D_n] \leqq P[C] - P[C \cup D_n] \leqq P[B_n] - P[B_n \cup D_n] \quad . \quad (5.6)$$

Since $\{X_n\}$ converges in probability to X,

$$\lim_{n \to \infty} P[D_n] = 1 \quad ,$$

and since

$$D_n \subset A_n \cup D_n \quad ,$$

we have

$$P[A_n \cup D_n] = 1 - \eta$$

where η is an infinitesimal as n increases without limit. Similarly

$$P[C \cup D_n] = 1 - \zeta$$

and

$$P[B_n \cup D_n] = 1 - \theta$$

where ζ and θ are also infinitesimals as n increases without limit.

We may therefore write (5.6) as

$$F_n(x - \varepsilon) + \eta \leqq G(x) + \zeta \leqq F_n(x + \varepsilon) + \theta \quad .$$

Since $\{F_n\}$ converges to F, we may take the limit to obtain

$$F(x - \varepsilon) \leqq G(x) \leqq F(x + \varepsilon) \quad .$$

Thus $F = G$ at every continuity point of F.

6. Convergence in Mean Square

Let $\{\xi_n\}$ be a sequence of real or complex random variables (P-measurable functions) defined on a probability space $\mathscr{P} = (\Omega, \mathscr{F}, P)$. Let ξ be a random variable defined on $\mathscr{P}$. If for every $\varepsilon > 0$ there exists an integer N such that

$$\mathscr{E} \,|\, \xi_n - \xi \,|^2 \leqq \varepsilon \tag{6.1}$$

for all $n > N$, then we say $\{\xi_n\}$ *converges in mean square to* ξ. Alternate terms are *convergence in the mean, convergence in quadratic mean.* By definition of limit we may write (6.1) as

$$\lim_{n \to \infty} \mathscr{E} \,|\, \xi_n - \xi \,|^2 = 0 \quad . \tag{6.2}$$

In terms of "limit" terminology

$$\lim_{n \to \infty} \xi_n = \xi \qquad [\text{mean sq}] \quad . \tag{6.3}$$

We prefer to drop the "mean sq" and write

$$\underset{n \to \infty}{\text{l.i.m.}}\ \xi_n = \xi \tag{6.4}$$

if indeed $\{\xi_n\}$ converges in mean square to ξ. [Equation (6.4) is read "limit in the mean."]

Let $\mathscr{H}$ be the set of all random variables η defined on $\mathscr{P}$ such that $\mathscr{E}\eta = 0$ and $\mathscr{E}|\eta|^2 < \infty$. Then $\mathscr{H}$ is a realization of (in general nonseparable) abstract Hilbert space. Clearly $\mathscr{H}$ is a linear vector space over the complex number field, and if we let $(\eta, \zeta) = \mathscr{E}\eta\bar{\zeta}$ for $\eta, \zeta \in \mathscr{H}$ be the inner product, then we easily see that $\mathscr{H}$ is unitary. The completeness of $\mathscr{H}$ is the property we need for Cauchy convergence. But $\mathscr{H}$ is complete (see [Mu, page 243]). Thus if $\{\eta_n\}$ is a Cauchy sequence which converges in mean square, that is, if for every positive ε there is an N such that

$$\mathscr{E}|\eta_n - \eta_m|^2 \leqq \varepsilon \tag{6.5}$$

for all $m, n > N$, then there exists a random variable $\eta \in \mathscr{H}$ such that $\{\eta_n\}$ converges in mean square to η.

In general a random variable ξ defined on $\mathscr{P}$ will not have mean zero. However, since $\mathscr{E}|\xi - \mathscr{E}\xi|^2 \geqq 0$ implies $|\mathscr{E}\xi|^2 \leqq \mathscr{E}|\xi|^2$, we see that if $\mathscr{E}|\xi|^2 < \infty$, then $\mathscr{E}\xi$ is finite. Now let $\{\xi_n\}$ be a sequence of random variables defined on $\mathscr{P}$, and let $\mathscr{E}|\xi_n|^2 < \infty$ for all n. Then $\{\eta_n\}$ where $\eta_n = \xi_n - \mathscr{E}\xi_n$ is defined on $\mathscr{H}$.

Suppose $\{\xi_n\}$ is a Cauchy sequence which converges in mean square. Then for any $\varepsilon > 0$ there exists an integer N such that

$$\mathscr{E}|\xi_n - \xi_m|^2 \leqq \varepsilon$$

for all $n, m > N$. But from the identity

$$\mathscr{E}|\xi_n - \xi_m|^2 = \mathscr{E}|\eta_n - \eta_m|^2 + |\mathscr{E}\xi_n - \mathscr{E}\xi_m|^2$$

we conclude that

$$\mathscr{E}|\eta_n - \eta_m|^2 \leqq \varepsilon \tag{6.6}$$

and

$$|\mathscr{E}\xi_n - \mathscr{E}\xi_m| \leqq \sqrt{\varepsilon} \tag{6.7}$$

for all $n, m > N$.

By (6.6), $\{\eta_n\}$ is a Cauchy sequence in $\mathscr{H}$ which converges in mean square. Hence there exists an $\eta \in \mathscr{H}$ such that $\{\eta_n\}$ converges in mean square to η. By (6.7) the sequence of complex numbers $\{\mathscr{E}\xi_n\}$ is also a Cauchy sequence and by the Cauchy convergence theorem for complex numbers, $\{\mathscr{E}\xi_n\}$ converges to some complex number, say c. Now let

$$\xi = \eta + c \quad .$$

Then $\{\xi_n\}$ converges in mean square to ξ.

We have thus shown that if $\lim_{n,m\to\infty}\mathscr{E}|\xi_n - \xi_m|^2 = 0$, then there exists a random variable ξ in $\mathscr{P}$, with finite second moment, such that $\lim_{n\to\infty} \mathscr{E}|\xi - \xi_n|^2 = 0$. By the triangle inequality

$$(\mathscr{E}|\xi_n - \xi_m|^2)^{1/2} = [\mathscr{E}|(\xi_n - \xi) + (\xi - \xi_m)|^2]^{1/2}$$

$$\leqq (\mathscr{E}|\xi_n - \xi|^2)^{1/2} + (\mathscr{E}|\xi - \xi_m|^2)^{1/2} \quad .$$

Thus we have the converse proposition that if $\lim_{n\to\infty} \mathscr{E}|\xi - \xi_n|^2 = 0$, then $\lim_{n,\, m\to\infty} \mathscr{E}|\xi_n - \xi_m|^2 = 0$. We also note that if $\{\xi_n\}$ converges in mean square to ξ, then by the above triangle inequality, $\{\xi_n\}$ is a Cauchy sequence which converges in mean square to ξ, and hence ξ has a finite second moment. Summarizing:

Theorem 6.1 (Cauchy Criterion). Let $\{\xi_n\}$ be a sequence of random variables defined on a probability space $\mathscr{P}$. Let $\mathscr{E}|\xi_n|^2 < \infty$ for all n. Then a necessary and sufficient condition that there exist a random variable ξ, with finite second moment, defined on $\mathscr{P}$ such that $\{\xi_n\}$ converge in mean square to ξ, is that

$$\lim_{n,m\to\infty} \mathscr{E}|\xi_n - \xi_m|^2 = 0 \quad .$$

An alternate criterion due to Loève that follows from the Cauchy criterion is given in Theorem 6.2.

Theorem 6.2 (Loève Criterion). Let $\{\xi_n\}$ be a sequence of random variables defined on a probability space $\mathscr{P}$. Let $\mathscr{E}|\xi_n|^2 < \infty$ for all n. Then a necessary and sufficient condition that there exist a random variable ξ, with finite second moment, defined on $\mathscr{P}$ such that $\{\xi_n\}$ converge in mean square to ξ, is that

$$\lim_{n,m\to\infty} \mathscr{E}\xi_n\bar{\xi}_m$$

approach a finite limit.

Proof. Suppose $\{\xi_n\}$ converges in mean square to ξ. Then by the Schwarz inequality

$$|\mathscr{E}(\xi - \xi_n)\,\bar{\xi}|^2 \leqq \mathscr{E}|\xi - \xi_n|^2\,\mathscr{E}|\xi|^2$$

and

$$\lim_{n\to\infty} \mathscr{E}\xi_n\bar{\xi} = \mathscr{E}|\xi|^2 = \mathscr{E}[\operatorname*{l.i.m.}_{n\to\infty}\, \xi_n\bar{\xi}] \quad . \tag{6.8}$$

Also

$$|\mathscr{E}(\xi - \xi_n)(\bar{\xi} - \bar{\xi}_m)|^2 \leqq \mathscr{E}|\xi - \xi_n|^2 \mathscr{E}|\xi - \xi_m|^2$$

together with (6.8) imply that

$$\lim_{n,m\to\infty} \mathscr{E}\xi_n\bar{\xi}_m = \mathscr{E}|\xi|^2 < \infty \quad .$$

Conversely, suppose $\mathscr{E}\xi_n\bar{\xi}_m$ approaches a finite limit, say c^2, as n and m independently increase without limit. Then

$$\mathscr{E}|\xi_n - \xi_m|^2 = \mathscr{E}|\xi_n|^2 - \mathscr{E}\xi_n\bar{\xi}_m - \mathscr{E}\bar{\xi}_n\xi_m + \mathscr{E}|\xi_m|^2$$

implies

$$\lim_{n,m\to\infty} \mathscr{E}|\xi_n - \xi_m|^2 = c^2 - c^2 - c^2 + c^2 = 0 \quad .$$

The Cauchy theorem now establishes the existence of a random variable ξ, with finite second moment, in the probability space $\mathscr{P}$ such that $\{\xi_n\}$ converges in mean square to ξ.

Corollary 6.1. If $\{\xi_n\}$ converges in mean square to ξ, then

$$\lim_{n\to\infty} \mathscr{E}\xi_n = \mathscr{E}\xi \tag{6.9}$$

and

$$\lim_{n,m\to\infty} \mathscr{E}\xi_n\bar{\xi}_m = \mathscr{E}|\xi|^2 \quad . \tag{6.10}$$

Proof. Since $|\mathscr{E}(\xi - \xi_n)|^2 \leqq \mathscr{E}|\xi - \xi_n|^2$ for any random variable $\xi - \xi_n$, we have

$$\mathscr{E}\xi = \lim_{n\to\infty} \mathscr{E}\xi_n = \mathscr{E}[\underset{n\to\infty}{\text{l.i.m.}}\ \xi_n] \quad .$$

Equation (6.10) is an immediate consequence of Theorem 6.2.

If $\{\alpha_n\}$ and $\{\beta_n\}$ are sequences of complex numbers which converge to α and β, respectively, then we know that $\{\alpha_n\beta_m\}$ converges to $\alpha\beta$. The corresponding result for random variables is given in the next theorem.

Theorem 6.3. Let $\{\xi_n\}$ and $\{\eta_n\}$ be sequences of random variables defined on the same probability space. Let $\mathscr{E}|\xi_n|^2 < \infty$ and $\mathscr{E}|\eta_n|^2 < \infty$ for all n. If $\{\xi_n\}$ converges in mean square to ξ and $\{\eta_n\}$ converges in mean square to η, then

$$\lim_{n,m\to\infty} \mathscr{E}\xi_n\bar{\eta}_m = \mathscr{E}\xi\bar{\eta} \quad .$$

Proof. By the Schwarz inequality

$$|\mathscr{E}(\xi - \xi_n)\,\bar{\eta}|^2 \leqq \mathscr{E}|\xi - \xi_n|^2\,\mathscr{E}|\eta|^2$$

and hence

$$\lim_{n\to\infty} \mathscr{E}\xi_n\bar{\eta} = \mathscr{E}\xi\bar{\eta} \quad .$$

Also by the Schwarz inequality

$$|\mathscr{E}(\xi - \xi_n)(\bar{\eta} - \bar{\eta}_m)|^2 \leqq \mathscr{E}|\xi - \xi_n|^2\,\mathscr{E}|\eta - \eta_m|^2$$

and hence

$$\lim_{n,m\to\infty} \mathscr{E}\xi_n\bar{\eta}_m = \mathscr{E}\xi\bar{\eta} \quad .$$

Let $\{\xi_n\}$ be a sequence of random variables defined on a probability space, and let $\{g_n\}$ be a sequence of complex numbers. We wish to investigate conditions under which the stochastic series

$$\sum_{n=1}^{\infty} g_n\xi_n \tag{6.11}$$

converges. By analogy with series of ordinary functions and the definition of mean square convergence, we shall say (6.11) *converges in mean square* if the sequence of partial sums $\{\sigma_n\}$, where

$$\sigma_n = \sum_{k=1}^{n} g_k\xi_k \quad ,$$

converges in mean square. A sufficient condition for mean square convergence is given in Theorem 6.4.

Theorem 6.4. Let $\{\xi_n\}$ be a sequence of random variables defined on a probability space $\mathscr{P}$. Let $\mathscr{E}|\xi_n|^2 < \infty$ for all n. Let $\{g_n\}$ be a sequence of complex numbers. Then a sufficient condition that (6.11) converge in mean square is that the series of real numbers

$$\sum_{k=1}^{\infty} |g_k|\,(\mathscr{E}|\xi_k|^2)^{1/2} \tag{6.12}$$

converge.

Proof. Let σ_n be the nth partial sum of (6.11). Then for $p > 0$

$$\sigma_{n+p} - \sigma_n = \sum_{k=n+1}^{n+p} g_k\xi_k$$

and

$$|\sigma_{n+p} - \sigma_n|^2 = \sum_{k=n+1}^{n+p} \sum_{j=n+1}^{n+p} g_k \bar{g}_j \xi_k \bar{\xi}_j \quad . \tag{6.13}$$

By the Schwarz inequality

$$|\mathscr{E}\xi_k\bar{\xi}_j| \leqq (\mathscr{E}|\xi_k|^2)^{1/2} (\mathscr{E}|\xi_j|^2)^{1/2} \quad .$$

Hence from (6.13)

$$\mathscr{E}|\sigma_{n+p} - \sigma_n|^2 \leqq \left[\sum_{k=n+1}^{n+p} |g_k| (\mathscr{E}|\xi_k|^2)^{1/2}\right]^2 \quad .$$

The hypothesis that (6.12) converge and Theorem 6.1 complete the proof.

Immediate consequences of this theorem are given in the three corollaries that follow. Each succeeding result is a weaker but simpler condition guaranteeing convergence.

Corollary 6.2. If $\mathscr{E}\xi_n\bar{\xi}_m = 0$ for $n \neq m$, then a sufficient condition that (6.11) converge is that

$$\sum_{n=1}^{\infty} |g_n|^2 \mathscr{E}|\xi_n|^2 < \infty \quad .$$

Corollary 6.3. If $\mathscr{E}|\xi_n|^2$ is independent of n for all n, then a sufficient condition that (6.11) converge is that

$$\sum_{n=1}^{\infty} |g_n| < \infty \quad .$$

Corollary 6.4. If $\mathscr{E}\xi_n\bar{\xi}_m = 0$ for $n \neq m$, and if $\mathscr{E}|\xi_n|^2$ is independent of n for all n, then a sufficient condition that (6.11) converge is that

$$\sum_{n=1}^{\infty} |g_n|^2 < \infty \quad .$$

Let $\{\xi_n\}$ be a sequence of random variables defined on a probability space $\mathscr{P}$. Let $\mathscr{E}|\xi_n|^2 < \infty$ for all n, and let $\{g_n\}$ be a sequence of complex numbers such that

$$\sum_{k=1}^{\infty} |g_k| \, (\mathscr{E}|\xi_k|^2)^{1/2} < \infty \quad .$$

Then by Theorem 6.4, the series $\sum_{k=1}^{\infty} g_k\xi_k$ converges in mean square (say to σ),

$$\sigma = \sum_{k=1}^{\infty} g_k \xi_k \quad . \tag{6.14}$$

Hence by Corollary 6.1

$$\mathscr{E}\sigma = \sum_{k=1}^{\infty} g_k \mathscr{E} \xi_k \quad ,$$

and since $\mathscr{E}|\sigma_n|^2 < \infty$ where σ_n is the nth partial sum of (6.14),

$$\mathscr{E}|\sigma|^2 = \sum_{k=1}^{\infty} \sum_{j=1}^{\infty} g_k \bar{g}_j \mathscr{E} \xi_k \bar{\xi}_j \quad .$$

If, in particular, $\mathscr{E}\xi_n\bar{\xi}_m = 0$ for $n \neq m$, then

$$\mathscr{E}|\sigma|^2 = \sum_{k=1}^{\infty} |g_k|^2 \mathscr{E}|\xi_k|^2 \quad .$$

Also, if $\{\eta_n\}$ is a sequence of random variables defined on the same probability space $\mathscr{P}$ with $\mathscr{E}|\eta_n|^2 < \infty$ for all n, and if $\{h_n\}$ is a sequence of complex numbers such that

$$\sum_{k=1}^{\infty} |h_k| \, (\mathscr{E}|\eta_k|^2)^{1/2} < \infty \quad ,$$

then the stochastic series $\sum_{k=1}^{\infty} h_k \eta_k$ converges in mean square (say to τ),

$$\tau = \sum_{k=1}^{\infty} h_k \eta_k \quad .$$

Thus by Theorem 6.3,

$$\mathscr{E}\sigma\bar{\tau} = \sum_{k=1}^{\infty} \sum_{j=1}^{\infty} g_k \bar{h}_j \mathscr{E} \xi_k \bar{\eta}_j \quad .$$

7. Some Examples of Convergence

In the three previous sections we have considered three different modes of convergence. Briefly, if $\{\xi_n\}$ is a sequence of random variables defined on a probability space $\mathscr{P}$, then if for every $\varepsilon > 0$

$$\lim_{N\to\infty} P[\bigcup_{n=N}^{\infty} \{|\xi_n - \xi| > \varepsilon\}] = 0 \quad ,$$

we say $\{\xi_n\}$ converges with probability one to ξ, and if

$$\lim_{N\to\infty} P[|\xi_N - \xi| > \varepsilon] = 0 \quad ,$$

we say $\{\xi_n\}$ converges in probability to ξ, while if

$$\lim_{N\to\infty} \mathscr{E}|\xi_N - \xi|^2 = 0 \quad ,$$

we say $\{\xi_n\}$ converges in mean square to ξ.

From (4.12) we have

$$P[|\xi_N - \xi| > \varepsilon] \leqq P[\bigcup_{n=N}^{\infty} \{|\xi_N - \xi| > \varepsilon\}] \tag{7.1}$$

and from (4.14) we have

$$P[|\xi_N - \xi| > \varepsilon] \leqq (1/\varepsilon^2)\, \mathscr{E}|\xi_N - \xi|^2 \quad . \tag{7.2}$$

Thus we conclude from (7.1) that if $\{\xi_n\}$ converges with probability one to ξ, then it converges in probability to ξ; and from (7.2) we conclude that if $\{\xi_n\}$ converges in mean square to ξ, then it converges in probability to ξ.

We now wish to give some counterexamples. First we shall show that convergence with probability one does not necessarily imply mean square convergence.

Example 7.1. Let $x_n = \xi_n - \xi$ where

$$P[x_n = n] = \frac{1}{n^2}$$

$$P[x_n = 0] = 1 - \frac{1}{n^2}$$

for all n. Then from (4.4), (4.3), and (4.12), for any $\varepsilon > 0$,

$$P[x_n > \varepsilon \quad \text{for at least one} \quad n \geqq N]$$

$$\leqq \sum_{n=N}^{\infty} P[x_n > \varepsilon] = \sum_{n=N}^{\infty} \frac{1}{n^2} \leqq \frac{\pi^2}{6}$$

and $\{\xi_n\}$ converges with probability one to ξ by (4.13). But since

$$\mathscr{E}x_n^2 = n^2\left(\frac{1}{n^2}\right) + 0\cdot\left(1 - \frac{1}{n^2}\right) = 1 \tag{7.3}$$

for all n, we see that $\{\xi_n\}$ does not converge in mean square to ξ.

[Note that

$$P[x_n > \varepsilon] = \frac{1}{n^2}$$

for all n implies that $\{\xi_n\}$ converges in probability to ξ by (5.3).]

As our second example we shall show that neither convergence in probability nor convergence in mean square necessarily imply convergence with probability one.

Example 7.2. Let $y_n = \xi_n - \xi$ where

$$P[y_n = 1] = \frac{1}{n}$$

$$P[y_n = 0] = 1 - \frac{1}{n}$$

for all n. Furthermore, let the y_n be independent for $n = 1, 2, \cdots$. Then for any $\varepsilon > 0$,

$$P[y_n > \varepsilon] = \frac{1}{n}$$

for all n, and $\{\xi_n\}$ converges in probability to ξ. Also since

$$\mathscr{E} y_n^2 = 1 \cdot \left(\frac{1}{n}\right) + 0 \cdot \left(1 - \frac{1}{n}\right) = \frac{1}{n}$$

for all n, we see that $\{\xi_n\}$ converges in mean square to ξ.

Now for any $\varepsilon > 0$,

$$P[y_n \leqq \varepsilon \quad \text{for all} \quad n \geqq N] = \lim_{m \to \infty} \prod_{k=N}^{m} \left(1 - \frac{1}{k}\right)$$

$$= \lim_{m \to \infty} \frac{N-1}{m} = 0$$

for any N. Hence [cf. (4.6)], we see that $\{\xi_n\}$ does not converge with probability one to ξ.

However, in spite of the negative results of Examples 7.1 and 7.2 we do have the results that convergence in mean square and convergence in probability each imply convergence with probability one for some subsequence.

Theorem 7.1. Let $\{\xi_n\}$ be a sequence of random variables, defined on a probability space $\mathscr{P}$, which converges in mean square to ξ. Then there exists a subsequence of $\{\xi_n\}$ which converges with probability one to ξ.

Proof. Since $\{\xi_n\}$ converges in mean square to ξ by hypothesis, we must have

$$\lim_{n \to \infty} \sigma_n^2 = 0$$

where

$$\sigma_n^2 = \mathscr{E}|\xi_n - \xi|^2 \quad .$$

Thus $\{\sigma_n^2\}$ is a sequence of nonnegative numbers which converges to zero.
Now choose an n_1 such that $\sigma_{n_1}^2 < \frac{1}{2}$, and an $n_k > n_{k-1}$ such that $\sigma_{n_k}^2 < 2^{-k}$ for $k = 2, 3, \cdots$. Then $\{\sigma_{n_k}^2\}$ converges to zero and

$$\sum_{k=1}^{\infty} \sigma_{n_k}^2 < 1 \quad .$$

Thus

$$\sum_{k=1}^{\infty} \mathscr{E}|\xi_{n_k} - \xi|^2 < \infty$$

and by (4.15) the subsequence $\{\xi_{n_k}\}$ converges with probability one to ξ.

Suppose the sequence $\{\xi_n\}$ also converges with probability one, say to ζ. Then Theorem 7.1 implies that ξ and ζ must be equivalent [cf. (3.5)]. Formally stated:

Corollary 7.1. If a sequence of random variables defined on some probability space converges in mean square and with probability one, then the limits are equal with probability one.

Now let us show that convergence in probability also implies convergence with probability one for some subsequence.

Theorem 7.2. Let $\{\xi_n\}$ be a sequence of random variables, defined on a probability space $\mathscr{P}$, which converges in probability to ξ. Then there exists a subsequence of $\{\xi_n\}$ which converges with probability one to ξ.

Proof. For any $\varepsilon > 0$ let

$$P[|\xi_n - \xi| > \varepsilon] = \delta_n \quad .$$

Then if $\{\xi_n\}$ converges in probability to ξ, we must have

$$\lim_{n\to\infty} \delta_n = 0 \quad .$$

Thus $\{\delta_n\}$ is a sequence of nonnegative numbers which converges to zero.
Now choose an n_1 such that $\delta_{n_1} < \frac{1}{2}$, and an $n_k > n_{k-1}$ such that $\delta_{n_k} < 2^{-k}$ for $k = 2, 3, \cdots$. Then $\{\delta_{n_k}\}$ converges to zero and

$$\sum_{k=1}^{\infty} P[|\xi_{n_k} - \xi| > \varepsilon] = \sum_{k=1}^{\infty} \delta_{n_k} < 1 < \infty \quad .$$

Thus by (4.13) the subsequence $\{\xi_{n_k}\}$ converges with probability one to ξ.

From (7.3) we see that if a sequence of random variables converges in probability, there does not necessarily exist a subsequence which converges in mean square.

II

CONTINUITY OF RANDOM VARIABLES

0. Introduction

Let $\xi(t, \omega)$ be a random variable for every t in some parameter set T. If we fix an elementary event $\omega = \omega_0$, then $\xi(t, \omega_0)$ is a *sample function.* This realization of the stochastic process is no longer a random variable. Thus we are invited to consider analytical properties of $\xi(t, \omega_0)$, for example, the continuity of ξ as a function of t. One can also "average" over all ω and consider the analytical properties of the "averaged" sample function.

The appropriate setting for these discussions is found in the theory of stochastic processes. We therefore begin by defining stochastic processes, equivalent processes, and stationary processes. A convenient tool for analyzing any stochastic process is its covariance function; and if the process is Gaussian, a knowledge of the mean and covariance function tells the whole story. If a process is stationary to the second order, the covariance function is a function of one variable only. In particular, with every continuous covariance function of a wide-sense stationary process, we may associate a spectral distribution function (Bochner's theorem). These preliminaries are disposed of in the first five sections.

In Section 6 we consider the continuity of $\xi(t, \omega)$ for almost all ω, and in Section 7 we consider the mean square continuity of a stochastic process. For weakly stationary processes, sufficient conditions for almost everywhere continuity or mean square continuity may be expressed in terms of the covariance function. By Bochner's theorem these conditions may be transferred to the spectral distribution function. Numerous theorems of this type are also proved in Sections 6 and 7. Finally, in Section 8, theorems similar to those proved in Sections 6 and 7 for continuity are established for differentiability almost everywhere and for differentiability in mean square.

1. Stochastic Processes

Let T be a subset of the real numbers. For every $t \in T$ let $x(t, \omega)$ be a random variable defined on a probability space $\mathscr{P} = (\Omega, \mathscr{F}, P)$. Then $\{x(t, \omega) \mid t \in T\}$ is called a (real) *stochastic process.* We call T a *linear index set* or a *parameter set.* Usually T is an interval (finite or infinite) or the set of positive integers. Thus, for example, the sequences of random variables considered in the previous chapter were stochastic processes. If $\{y(t, \omega) \mid t \in T\}$ is another (real) stochastic process defined on the same probability space $\mathscr{P}$, then $\{\xi(t, \omega) \mid t \in T\}$ where $\xi(t, \omega) = x(t, \omega) + iy(t, \omega)$ is a (complex) stochastic process. Thus $\mathscr{X} = \{\xi(t, \omega) \mid t \in T\}$ is a *complex stochastic process* if its real and imaginary parts are P-measurable functions on $\mathscr{P}$ for all $t \in T$. It is customary to drop the ω and write $\mathscr{X}$ simply as $\{\xi(t) \mid t \in T\}$. If $\{\xi(t) \mid t \in T\}$ and $\{\zeta(t) \mid t \in T\}$ are two stochastic processes defined on the same probability space $\mathscr{P}$ and the same parameter set T, then we say the processes are *equivalent* if for every $t_0 \in T$, the random variables $\xi(t_0)$ and $\zeta(t_0)$ are equivalent [see (3.5) of Chapter I.]

Let T be a linear index set. Suppose $K(t, s)$ is a function defined on $T \times T$ with the properties that

$$K(t, s) = \bar{K}(s, t) \quad , \qquad t, s \in T \quad , \tag{1.1}$$

and

$$\sum_{j,k=1}^{n} K(t_j, t_k)\lambda_j\bar{\lambda}_k \geqq 0 \tag{1.2}$$

for all complex numbers λ_j, $1 \leqq j \leqq n$, all sets of distinct points t_j, $1 \leqq j \leqq n$, in T, and all positive integers n. Then we say $K(t, s)$ is *nonnegative definite.* If the left-hand side of (1.2) vanishes if and only if all the λ_j, $1 \leqq j \leqq n$, are zero, then we say $K(t, s)$ is *positive definite.* A function $K(t)$ of one variable with domain T is said to be *nonnegative definite* if $K(t) = \bar{K}(-t)$ for all $t \in T$, and if

$$\sum_{j,k=1}^{n} K(t_j - t_k)\lambda_j\bar{\lambda}_k \geqq 0 \tag{1.3}$$

for all complex numbers λ_j, $1 \leqq j \leqq n$, all sets of distinct points t_j, $1 \leqq j \leqq n$, in T, and all positive integers n. (Cf. Section 3 of Chapter I.) If the left-hand side of (1.3) vanishes only when all the λ_j, $1 \leqq j \leqq n$, are zero, then we say $K(t)$ is *positive definite.*

Now consider a stochastic process $\{\xi(t) \mid t \in T\}$ with mean function $c(t)$,

$$\mathscr{E}\xi(t) = c(t) \quad ,$$

and covariance function $R(t, s)$,

$$\text{Cov}\,(\xi(t), \bar{\xi}(s)) = \mathscr{E}[\xi(t) - c(t)]\,[\bar{\xi}(s) - \bar{c}(s)] = R(t, s)$$

for all t, s in T. Clearly $R(t, s) = \bar{R}(s, t)$. Furthermore, if λ_j, $1 \leqq j \leqq n$, are arbitrary complex numbers, and if t_j, $1 \leqq j \leqq n$, are distinct points in T, then for all n,

$$\mathscr{E}\left|\sum_{j=1}^{n} \lambda_j[\xi(t_j) - c(t_j)]\right|^2 = \sum_{j,k=1}^{n} R(t_j, t_k)\lambda_j\bar{\lambda}_k \geqq 0 \quad .$$

Thus we see that any covariance function is nonnegative definite.

An important special situation occurs when the covariance function $R(t, s)$ depends only on the *difference* of its arguments. In this case we say the stochastic process is stationary. More precisely, let $\{\xi(t)\,|\,t \in T\}$ be a complex stochastic process. Let $\mathscr{E}\xi(t)$ be independent of t, and let $\mathscr{E}\,|\xi(t)|^2 < \infty$ for all $t \in T$. Then if $\text{Cov}(\xi(t), \bar{\xi}(s))$ depends only on the difference $t - s$, that is, if $\text{Cov}(\xi(t + h), \bar{\xi}(t)) = R(h)$, then we say $\{\xi(t)\,|\,t \in T\}$ is *stationary in the wide sense,* or *weakly stationary,* or *stationary to the second order.*

Let $\{\xi(t)\,|\,t \in T\}$ be a complex stochastic process. By the joint n-dimensional distribution of $\xi(t_1), \cdots, \xi(t_n)$ where the t_j, $1 \leqq j \leqq n$, are distinct points in T, we shall mean the joint $2n$-dimensional distribution of the real and imaginary components of $\xi(t_j)$, $1 \leqq j \leqq n$. (See Section 3 of Chapter I.) Now suppose that the joint distribution of $\xi(t_1), \cdots, \xi(t_n)$ is identical with the joint distribution of $\xi(t_1 + h), \cdots, \xi(t_n + h)$ for all $t_1, \cdots, t_n$ in T and all h such that $t_1 + h, \cdots, t_n + h \in T$, and all n. Then we say $\{\xi(t)\,|\,t \in T\}$ is *stationary in the strict sense.* Clearly, for a strictly stationary process, $\mathscr{E}\xi(t) = \mathscr{E}\xi(s)$ for all $t, s \in T$. Thus, if a strictly stationary process has finite second moments, it is weakly stationary. The converse is not necessarily true.

2. Hermitian Matrices

Closely related to the notion of positive definite functions are the concepts of positive definite matrices and positive definite quadratic forms. In this section we shall give the pertinent definitions together with some of the elementary properties and identities involving positive definite matrices. The relation between a positive definite matrix and the covariance function of a random vector will be established.

Let $\boldsymbol{H} = \|\,H_{jk}\,\|$ be an $n \times n$ square matrix where the H_{jk} are complex numbers. We call $\bar{\boldsymbol{H}} = \|\,\bar{H}_{jk}\,\|$ the *conjugate* of $\boldsymbol{H}$. If $\boldsymbol{H}' = \bar{\boldsymbol{H}}$ where $\boldsymbol{H}'$ is the transpose of $\boldsymbol{H}$, then $\boldsymbol{H}$ is called a *Hermitian matrix.* Thus, for a Hermitian matrix, $H_{jk} = \bar{H}_{kj}$, $1 \leqq j, k \leqq n$. We may write $\boldsymbol{H}$ in terms of real and imaginary parts, viz., $\boldsymbol{H} = \text{Re}\,\boldsymbol{H} + i\,\text{Im}\,\boldsymbol{H}$. Then Re $\boldsymbol{H}$ is a real symmetric matrix and Im $\boldsymbol{H}$ is a real skew-symmetric matrix. The sum of two Hermitian matrices is again Hermitian; the product of two Hermitian matrices is Hermitian if and only if the matrices commute. If $\boldsymbol{H}$ is nonsingular, its inverse is also Hermitian. All the characteristic zeros of a Hermitian matrix are real.

A complex $n \times n$ matrix $\boldsymbol{K}$ is called *skew-Hermitian* if $\boldsymbol{K}' = -\bar{\boldsymbol{K}}$. If $\boldsymbol{\Psi}$ is any $n \times n$ complex matrix, then $\boldsymbol{\Psi} + \bar{\boldsymbol{\Psi}}'$ is Hermitian and $\boldsymbol{\Psi} - \bar{\boldsymbol{\Psi}}'$ is skew-Hermitian. Hence any $n \times n$ complex matrix $\boldsymbol{\Psi}$ may be written in the form

$$\boldsymbol{\Psi} = \boldsymbol{H} + \boldsymbol{K}$$

where $\boldsymbol{H}$ is Hermitian and $\boldsymbol{K}$ is skew-Hermitian.

Let $\boldsymbol{z}' = \{z_1, \cdots, z_n\}$ be a (complex) row vector; that is, we allow the z_j, $1 \leqq j \leqq n$, to be complex numbers. We call $\bar{\boldsymbol{z}}' = \{\bar{z}_1, \cdots, \bar{z}_n\}$ the conjugate of $\boldsymbol{z}'$. Let $z_j = x_j + iy_j$ where $x_j = \text{Re } z_j$, $y_j = \text{Im } z_j$, $1 \leqq j \leqq n$. Then we may write $\boldsymbol{z}' = \boldsymbol{x}' + i\boldsymbol{y}'$ where $\boldsymbol{x}' = \{x_1, \cdots, x_n\} = \text{Re } \boldsymbol{z}'$ and $\boldsymbol{y}' = \{y_1, \cdots, y_n\} = \text{Im } \boldsymbol{z}'$.

Now let $\boldsymbol{H}$ be an $n \times n$ Hermitian matrix and let $\boldsymbol{z}'$ be an n-dimensional complex vector. We wish to consider the quadratic form

$$q = \bar{\boldsymbol{z}}'\boldsymbol{H}\boldsymbol{z} \quad .$$

(Primed vectors will always denote row vectors, and hence unprimed (transposed) vectors are column vectors.) It is easily seen that q is real, for since q is a scalar

$$\bar{q} = \overline{\bar{\boldsymbol{z}}'\boldsymbol{H}\boldsymbol{z}} = \boldsymbol{z}'\bar{\boldsymbol{H}}\bar{\boldsymbol{z}} = \boldsymbol{z}'\boldsymbol{H}'\bar{\boldsymbol{z}} = (\boldsymbol{z}'\boldsymbol{H}'\bar{\boldsymbol{z}})' = \bar{\boldsymbol{z}}'\boldsymbol{H}\boldsymbol{z} = q \quad .$$

If $q \geqq 0$ for all complex vectors $\boldsymbol{z}$, then we call q a *nonnegative definite quadratic form* and we call $\boldsymbol{H}$ a *nonnegative definite matrix.* If $q \geqq 0$ for all $\boldsymbol{z}$ and $q = 0$ if and only if $\boldsymbol{z}$ is the zero vector, then we call q a *positive definite quadratic form* and we call $\boldsymbol{H}$ a *positive definite matrix.* All the characteristic zeros of a nonnegative definite matrix are nonnegative; all the characteristic zeros of a positive definite matrix are positive.

If the positive definite $n \times n$ Hermitian matrix $\boldsymbol{H}$ has the real, positive (but not necessarily distinct) characteristic zeros $\lambda_1, \cdots, \lambda_n$, then there exists a unitary matrix $\boldsymbol{U}$ such that

$$\bar{\boldsymbol{U}}'\boldsymbol{H}\boldsymbol{U} = \| \lambda_j \delta_{jk} \| \quad .$$

Hence there exists a nonsingular $n \times n$ matrix $\boldsymbol{Q}$ such that

$$\bar{\boldsymbol{Q}}'\boldsymbol{H}\boldsymbol{Q} = \boldsymbol{I}$$

where $\boldsymbol{I}$ is the $n \times n$ identity matrix. Thus

$$\boldsymbol{H} = \bar{\boldsymbol{P}}'\boldsymbol{P} \tag{2.1}$$

(where $\boldsymbol{P} = \boldsymbol{Q}^{-1}$). In fact, a Hermitian matrix $\boldsymbol{H}$ is positive definite if and only if there exists a nonsingular $n \times n$ matrix $\boldsymbol{P}$ such that (2.1) holds. We may therefore conclude that if $\boldsymbol{H}$ is positive definite, then so is its inverse. Furthermore, if $\boldsymbol{C}$ is any $n \times p$ matrix of rank p, and if $\boldsymbol{H}$ is an $n \times n$ positive definite matrix, then $\bar{\boldsymbol{C}}'\boldsymbol{H}\boldsymbol{C}$ is also positive definite. For if $\boldsymbol{z}$ is a nonzero p-dimensional vector, and we let $\boldsymbol{C}\boldsymbol{z} = \boldsymbol{\zeta}$, then since the rank of $\boldsymbol{C}$ is p, the n-dimensional vector $\boldsymbol{\zeta}$ is nonzero. Thus $\bar{\boldsymbol{z}}'(\bar{\boldsymbol{C}}'\boldsymbol{H}\boldsymbol{C})\boldsymbol{z} = (\bar{\boldsymbol{C}}\bar{\boldsymbol{z}})'\boldsymbol{H}(\boldsymbol{C}\boldsymbol{z}) = \bar{\boldsymbol{\zeta}}'\boldsymbol{H}\boldsymbol{\zeta}$, which is positive definite. The converse is also true, that is, $\bar{\boldsymbol{C}}'\boldsymbol{H}\boldsymbol{C}$ is positive definite only if $\boldsymbol{C}$ is of rank p.

We shall now deduce some elementary identities that will prove useful in later chapters. Toward this end, let $\boldsymbol{H}$ be an $n \times n$ Hermitian matrix and let $\boldsymbol{z}$ be an n-dimensional complex vector. Let

$$q = \bar{\boldsymbol{z}}'\boldsymbol{H}\boldsymbol{z}$$

be a (real) quadratic form. Then some simple matrix manipulations suffice to show that

$$q = \tfrac{1}{2}\boldsymbol{\eta}'\boldsymbol{F}\boldsymbol{\eta} = \tfrac{1}{2}\bar{\boldsymbol{\eta}}'\boldsymbol{G}\boldsymbol{\eta} = \tfrac{1}{2}\boldsymbol{\xi}'\boldsymbol{W}\boldsymbol{\xi} \tag{2.2}$$

where

$$\boldsymbol{\eta} = \begin{bmatrix} \boldsymbol{z} \\ \bar{\boldsymbol{z}} \end{bmatrix}$$

and

$$\boldsymbol{\xi} = \begin{bmatrix} \boldsymbol{x} \\ \boldsymbol{y} \end{bmatrix}$$

are $2n$-dimensional column vectors ($\bar{\boldsymbol{z}}$ being the conjugate of $\boldsymbol{z}$, and $\boldsymbol{x} = \operatorname{Re} \boldsymbol{z}$, $\boldsymbol{y} = \operatorname{Im} \boldsymbol{z}$). The $2n \times 2n$ matrices $\boldsymbol{F}$, $\boldsymbol{G}$, and $\boldsymbol{W}$ are given by

$$\boldsymbol{F} = \begin{bmatrix} \boldsymbol{0} & \bar{\boldsymbol{H}} \\ \boldsymbol{H} & \boldsymbol{0} \end{bmatrix},$$

$$\boldsymbol{G} = \begin{bmatrix} \boldsymbol{H} & \boldsymbol{0} \\ \boldsymbol{0} & \bar{\boldsymbol{H}} \end{bmatrix},$$

and

$$\boldsymbol{W} = 2 \operatorname{Re} \begin{bmatrix} \boldsymbol{H} & i\boldsymbol{H} \\ i\bar{\boldsymbol{H}} & \boldsymbol{H} \end{bmatrix} \tag{2.3}$$

where $\bar{\boldsymbol{H}}$ is the conjugate of $\boldsymbol{H}$ and $\boldsymbol{0}$ is the $n \times n$ zero matrix. Since $\boldsymbol{W}$ is real, (2.2) implies that if $\boldsymbol{H}$ is positive definite, then $\boldsymbol{W}$ is a real symmetric positive definite matrix.

If we assume $\boldsymbol{H}$ to be nonsingular, then $\boldsymbol{K} = \boldsymbol{H}^{-1}$ exists, and

$$F^{-1} = \begin{bmatrix} 0 & K \\ \bar{K} & 0 \end{bmatrix} \quad ,$$

$$G^{-1} = \begin{bmatrix} K & 0 \\ 0 & \bar{K} \end{bmatrix} \quad ,$$

and

$$W^{-1} = \tfrac{1}{2}\,\mathrm{Re}\begin{bmatrix} K & iK \\ i\bar{K} & K \end{bmatrix}$$

where $\bar{K}$ is the conjugate of K. We may also compute the determinants of F, G, and W, namely,

$$\det F = (-1)^n(\det H)^2 \quad ,$$

$$\det G = (\det H)^2 \quad ,$$

and

$$\det W = 2^{2n}(\det H)^2 \quad .$$

Let us now make the identification between Hermitian matrices and random vectors. Suppose, then, that $\{\zeta(t)\,|\,t\in T\}$ is a (complex) stochastic process with mean function $c(t)$ and covariance function $R(t, s)$. Let $t_1, \cdots, t_n$ be n distinct points in T. Then we are invited to consider the n-dimensional random vector

$$\zeta' = \{\zeta_1, \cdots, \zeta_n\}$$

where $\zeta_j = \zeta(t_j)$, $1 \leqq j \leqq n$. The *mean vector* c of ζ is

$$\mathscr{E}\zeta' = c' = \{c_1, \cdots, c_n\}$$

where $c_j = c(t_j)$, $1 \leqq j \leqq n$; and the *covariance matrix* R of ζ is

$$\mathscr{E}(\zeta - c)(\bar{\zeta} - \bar{c})' = R = \|\, R_{jk} \,\| \tag{2.4}$$

where $R_{jk} = R(t_j, t_k)$, $1 \leqq j, k \leqq n$. If θ is an n-dimensional complex vector, then

$$\bar{\theta}'R\theta = \mathscr{E}\,|\,\bar{\theta}'(\zeta - c)\,|^2 \geqq 0 \quad . \tag{2.5}$$

Thus R is nonnegative definite.

3. Real Gaussian Distributions

Let x be a real random variable with mean μ and variance σ^2. That is,

$$\mathscr{E}x = \mu$$

and

$$\operatorname{Var} x = \mathscr{E}(x - \mu)^2 = \sigma^2 \quad .$$

If the density function of x is

$$f(x) = \frac{1}{\sigma\sqrt{2\pi}} \exp\left(-\frac{(x-\mu)^2}{2\sigma^2}\right)$$

then we say x is *normally distributed* (μ, σ^2). Clearly $f(x) \geqq 0$ for all x and

$$\int_{-\infty}^{\infty} f(x)\, dx = 1 \quad ,$$

so that $f(x)$ is indeed a density function.

Now let $\boldsymbol{x}' = \{x_1, \cdots, x_n\}$ be a real random vector with mean vector $\boldsymbol{\mu}'$ and positive definite covariance matrix $\boldsymbol{\Sigma}$. That is

$$\mathscr{E}\boldsymbol{x} = \boldsymbol{\mu}$$

and

$$\operatorname{Var} \boldsymbol{x} = \mathscr{E}(\boldsymbol{x} - \boldsymbol{\mu})(\boldsymbol{x} - \boldsymbol{\mu})' = \boldsymbol{\Sigma} \quad . \tag{3.1}$$

If the density function of $\boldsymbol{x}$ is

$$f(\boldsymbol{x}) = \frac{1}{(2\pi)^{n/2}(\det \boldsymbol{\Sigma})^{1/2}} \exp\left[-\tfrac{1}{2}(\boldsymbol{x} - \boldsymbol{\mu})'\boldsymbol{\Sigma}^{-1}(\boldsymbol{x} - \boldsymbol{\mu})\right] \tag{3.2}$$

then we say $\boldsymbol{x}$ is *normally distributed* $(\boldsymbol{\mu}, \boldsymbol{\Sigma})$ in n dimensions. Clearly $f(\boldsymbol{x}) \geqq 0$ over all of n-dimensional Euclidean space. Thus in order to verify that $f(\boldsymbol{x})$ is indeed a density function, it remains but to show that

$$\int_{-\infty}^{\infty} \cdots \int_{-\infty}^{\infty} f(\boldsymbol{x}) dx_1 \cdots dx_n \tag{3.3}$$

has the value unity. We shall now demonstrate this fact.

First, we shall write the n-dimensional integral (3.3) in the compact form

$$\int_{-\infty}^{\infty} f(\boldsymbol{x}) d\boldsymbol{x} \tag{3.4}$$

where $d\boldsymbol{x}$ is to be interpreted as $dx_1 \cdots dx_n$. Since $\boldsymbol{\Sigma}$ is positive definite, there exists a nonsingular real matrix $\boldsymbol{Q}$ such that $\boldsymbol{\Sigma} = \boldsymbol{Q}\boldsymbol{Q}'$. Now make the change of variable

$$\boldsymbol{x} = \boldsymbol{Q}\boldsymbol{y} + \boldsymbol{\mu}$$

with Jacobian

$$\left|\frac{\partial \boldsymbol{x}}{\partial \boldsymbol{y}}\right| = |\det \boldsymbol{Q}| \quad .$$

Then (3.4) becomes

$$\int_{-\infty}^{\infty} f(\boldsymbol{x})d\boldsymbol{x} = \frac{1}{(2\pi)^{n/2}|\det \boldsymbol{Q}|}\int_{-\infty}^{\infty} \exp\left(-\tfrac{1}{2}\boldsymbol{y}'\boldsymbol{y}\right)|\det \boldsymbol{Q}|\, d\boldsymbol{y}$$

$$= \frac{1}{(2\pi)^{n/2}}\int_{-\infty}^{\infty} \exp\left(-\tfrac{1}{2}\sum_{j=1}^{n} y_j^2\right)d\boldsymbol{y} = 1 \tag{3.5}$$

where $\boldsymbol{y}' = \{y_1, \cdots, y_n\}$.

An important property of Gaussian vectors is that their marginal density functions are also normal. Let p and q be nonnegative integers whose sum is n, and partition $\boldsymbol{x}$, $\boldsymbol{\mu}$, and $\boldsymbol{\Sigma}$ as

$$\boldsymbol{x} = \begin{bmatrix} \boldsymbol{x}_p \\ \boldsymbol{x}_q \end{bmatrix} , \quad \boldsymbol{\mu} = \begin{bmatrix} \boldsymbol{\mu}_p \\ \boldsymbol{\mu}_q \end{bmatrix} ,$$

and

$$\boldsymbol{\Sigma} = \begin{bmatrix} \boldsymbol{\Sigma}_{pp} & \boldsymbol{\Sigma}_{pq} \\ \boldsymbol{\Sigma}_{qp} & \boldsymbol{\Sigma}_{qq} \end{bmatrix} ,$$

where $\boldsymbol{x}_p$ is p-dimensional, $\boldsymbol{x}_q$ is q-dimensional, etc. Then the density function of $\boldsymbol{x}_p$ is

$$f_p(\boldsymbol{x}_p) = \int_{-\infty}^{\infty} f(\boldsymbol{x})d\boldsymbol{x}_q$$

$$= \frac{1}{(2\pi)^{p/2}(\det \boldsymbol{\Sigma}_{pp})^{1/2}} \exp\left[-\tfrac{1}{2}(\boldsymbol{x}_p - \boldsymbol{\mu}_p)'\,\boldsymbol{\Sigma}_{pp}^{-1}(\boldsymbol{x}_p - \boldsymbol{\mu}_p)\right] \quad . \tag{3.6}$$

Thus $\boldsymbol{x}_p$ is normally distributed $(\boldsymbol{\mu}_p, \boldsymbol{\Sigma}_{pp})$ in p dimensions. (For a proof of this result and other properties of real Gaussian vectors see [Mil 1].)

The moment generating function $m(\boldsymbol{t})$ of $\boldsymbol{x}$ is

$$m(\boldsymbol{t}) = \mathscr{E} e^{\boldsymbol{t}'\boldsymbol{x}}$$

where $\boldsymbol{t}$ is a real n-dimensional vector independent of $\boldsymbol{x}$. Thus

$$m(\boldsymbol{t}) = \int_{-\infty}^{\infty} e^{\boldsymbol{t}'\boldsymbol{x}} f(\boldsymbol{x})d\boldsymbol{x}$$

$$= \frac{1}{(2\pi)^{n/2}(\det \boldsymbol{\Sigma})^{1/2}}\int_{-\infty}^{\infty} \exp\left[\boldsymbol{t}'\boldsymbol{x} - \tfrac{1}{2}(\boldsymbol{x} - \boldsymbol{\mu})'\,\boldsymbol{\Sigma}^{-1}(\boldsymbol{x} - \boldsymbol{\mu})\right]d\boldsymbol{x} \quad .$$

But some elementary matrix manipulations show that

$$\begin{aligned} &\boldsymbol{t}'\boldsymbol{x} - \tfrac{1}{2}(\boldsymbol{x} - \boldsymbol{\mu})'\boldsymbol{\Sigma}^{-1}(\boldsymbol{x} - \boldsymbol{\mu}) \\ &\qquad = \tfrac{1}{2}\boldsymbol{t}'\boldsymbol{\Sigma}\boldsymbol{t} + \boldsymbol{t}'\boldsymbol{\mu} - \tfrac{1}{2}(\boldsymbol{x} - \boldsymbol{\mu} - \boldsymbol{\Sigma}\boldsymbol{t})'\boldsymbol{\Sigma}^{-1}(\boldsymbol{x} - \boldsymbol{\mu} - \boldsymbol{\Sigma}\boldsymbol{t}) \enspace . \end{aligned}$$

Hence

$$m(\boldsymbol{t}) = \exp\left(\tfrac{1}{2}\boldsymbol{t}'\boldsymbol{\Sigma}\boldsymbol{t} + \boldsymbol{t}'\boldsymbol{\mu}\right) \enspace . \tag{3.7}$$

The characteristic function ϕ of $\boldsymbol{x}$ is

$$\phi(\boldsymbol{t}) = \mathscr{E}\exp(i\boldsymbol{t}'\boldsymbol{x}) = \exp\left(-\tfrac{1}{2}\boldsymbol{t}'\boldsymbol{\Sigma}\boldsymbol{t} + i\boldsymbol{t}'\boldsymbol{\mu}\right) \tag{3.8}$$

where $\boldsymbol{t}$ is a real n-dimensional vector independent of $\boldsymbol{x}$.

Suppose now that $\{x(t)\,|\,t \in T\}$ is a *real* stochastic process. If the joint distribution of $x(t_1), \cdots, x(t_n)$ is n-dimensional Gaussian for any choice of distinct points $t_1, \cdots, t_n$ in T, and for any n, then we say $\{x(t)\,|\,t \in T\}$ is a *real Gaussian process*. Now let $\{\xi(t)\,|\,t \in T\}$ be a *complex* stochastic process, and let $\xi(t) = x(t) + iy(t)$ for all $t \in T$. If the joint distribution of $x(t_1)$, $y(t_1), \cdots, x(t_n)$, $y(t_n)$ is $2n$-dimensional Gaussian for any choice of distinct points $t_1, \cdots, t_n$ in T, and for any n, then we say $\{\xi(t)\,|\,t \in T\}$ is a *complex Gaussian process.*

4. Complex Gaussian Processes

Gaussian stochastic processes appear with amazing frequency in both theoretical and applied analysis. We shall consider some properties of such processes in the present section. Further discussion and applications of complex Gaussian processes will appear in later chapters.

We begin by considering the relationship between strictly stationary and wide-sense stationary Gaussian processes. Since all the moments of a Gaussian process exist, we conclude that a strictly stationary Gaussian process is weakly stationary. The converse is also true for *real* Gaussian processes (Theorem 4.1) and, with one additional restriction, for *complex* Gaussian processes (Theorem 4.2).

Theorem 4.1. Let $\{x(t)\,|\,t \in T\}$ be a wide-sense stationary real Gaussian process. Then $\{x(t)\,|\,t \in T\}$ is strictly stationary.

Proof. If $\{x(t)\,|\,t \in T\}$ is *any* real Gaussian process with mean $a(t) = \mathscr{E}x(t)$ and covariance function $M(t, s) = \text{Cov}\,(x(t), x(s))$, then the joint distribution of any finite set of variates $x(t_1), \cdots, x(t_n)$ (where the t_j, $1 \leqq j \leqq n$, are distinct points in T) depends only on $a(t)$ and $M(t, s)$. That is, if $\boldsymbol{x}' = \{x(t_1), \cdots, x(t_n)\}$, then

$$\mathscr{E}\boldsymbol{x}' = \{a(t_1), \cdots, a(t_n)\} = \boldsymbol{a}'$$

and

$$\mathscr{E}(\boldsymbol{x} - \boldsymbol{a})(\boldsymbol{x} - \boldsymbol{a})' = \| M(t_j, t_k) \|$$

[see (3.1)]. Thus to show that a real Gaussian process is strictly stationary, it is sufficient to prove that $a(t + h) = a(t)$ and $M(t + h, s + h) = M(t, s)$ for all $t, s, t + h, s + h$ in T.

Now let us utilize the hypothesis of weak stationarity. Then $\mathscr{E}x(t) = a$ independent of t, and

$$\mathrm{Cov}\,(x(t + h), x(s + h)) = M((t + h) - (s + h)) = M(t - s)$$

—which proves the theorem.

In the light of the theorem above we see that the term *stationary real Gaussian process* is unambiguous.

For *complex* Gaussian processes, the state of affairs is slightly more complicated. The key result is given in Theorem 4.2.

Theorem 4.2. Let $\{\xi(t) | t \in T\}$ be a wide-sense stationary complex Gaussian process. Then $\{\xi(t) | t \in T\}$ is stationary in the strict sense if and only if $\mathrm{Cov}\,(\xi(t), \xi(s))$ is a function only of $t - s$ for all $t, s \in T$.

Proof. Let $\mathscr{E}\xi(t) = a(t)$ and $\mathrm{Cov}\,(\xi(t), \bar{\xi}(s)) = R(t, s)$. Then if the process is weakly stationary, $\mathscr{E}\xi(t) = a$ independent of t and

$$R(t, s) = R(t - s)$$

for all $t, s \in T$.

Now to examine the *strict* stationarity of the $\{\xi(t) | t \in T\}$ process, we must consider the joint distribution of the real and imaginary components of $\xi(t)$. Since the process is Gaussian with constant mean, we need only examine the $2n$-dimensional covariance matrix of $x(t_1), y(t_1), \cdots, x(t_n), y(t_n)$ where $x(t) = \mathrm{Re}\, \xi(t)$, $y(t) = \mathrm{Im}\, \xi(t)$, and the $t_1, \cdots, t_n$ are distinct points in T. That is, we must show that $\mathrm{Cov}\,(x(t), x(s))$, $\mathrm{Cov}\,(y(t), y(s))$, $\mathrm{Cov}\,(x(t), y(s))$ all depend only on the difference $t - s$ for all $t, s \in T$. But this is equivalent to showing that $\mathrm{Cov}\,(\xi(t), \bar{\xi}(s))$ and $\mathrm{Cov}\,(\xi(t), \xi(s))$ are functions only of $t - s$. Clearly $\mathrm{Cov}\,(\xi(t), \bar{\xi}(s))$ depends only on the difference of its arguments since $\mathrm{Cov}\,(\xi(t), \bar{\xi}(s)) = R(t - s)$. However, there is no guarantee that, in general, $\mathrm{Cov}\,(\xi(t), \xi(s))$ will be a function only of $t - s$. Thus we conclude that a wide-sense stationary complex Gaussian process is strictly stationary if and only if $\mathrm{Cov}\,(\xi(t), \xi(s))$ is a function only of $t - s$ for all $t, s \in T$.

We have defined Gaussian processes $\{\xi(t)|t \in T\}$ in terms of the joint distribution of $\xi(t_1), \cdots, \xi(t_n)$, for arbitrary n, and distinct points t_j, $1 \leq j \leq n$, in T. This, of course, implies that the covariance matrix of $\xi(t_1), \cdots, \xi(t_n)$ is positive definite, or, equivalently, that the covariance function of the $\{\xi(t)|t \in T\}$ process is positive definite. The definition may be weakened. That is, we shall say $\{\xi(t)|t \in T\}$ is a Gaussian process if the characteristic function [see (3.8)] of $\{\xi(t_1), \cdots, \xi(t_n)\}$ exists. Since the characteristic fuction does not depend on the inverse covariance matrix, it will exist even if the covariance matrix is singular. Thus we may define Gaussian processes even if the covariance matrix (covariance function) is nonnegative definite, rather than under the more restrictive assumption that the covariance matrix (covariance function) be positive definite. Of course, in such a situation, we have a *singular* Gaussian distribution.

If two stochastic processes $\{\xi(t)|t \in T\}$ and $\{\eta(t)|t \in T\}$ have the property that $\{\xi(t_1), \cdots, \xi(t_n)\}$ and $\{\eta(t_1), \cdots, \eta(t_n)\}$ have the same joint density function for all sets of distinct points $t_1, \cdots, t_n$ in T and for all n, then we shall say that $\{\xi(t)|t \in T\}$ and $\{\eta(t)|t \in T\}$ are *identically distributed.* For example, if $\{x(t)|t \in T\}$ and $\{y(t)|t \in T\}$ are two *real* Gaussian processes defined on the same parameter set T, and if

$$\mathscr{E}x(t) = \mathscr{E}y(t) \quad , \qquad t \in T \quad ,$$

and

$$\text{Cov}\,(x(t), x(s)) = \text{Cov}\,(y(t), y(s)) \quad , \qquad t, s \in T \quad ,$$

then it is easily seen that $\{x(t)|t \in T\}$ and $\{y(t)|t \in T\}$ are identically distributed.

Suppose now that $\{\xi(t)|t \in T\}$ and $\{\eta(t)|t \in T\}$ are two *complex* Gaussian processes defined on the same parameter set T. Let

$$\mathscr{E}\xi(t) = \mathscr{E}\eta(t) \quad , \qquad t \in T \quad ,$$

and

$$\text{Cov}\,(\xi(t), \bar{\xi}(s)) = \text{Cov}\,(\eta(t), \bar{\eta}(s)) \quad , \qquad t, s \in T \quad .$$

Then it does not follow that $\{\xi(t)|t \in T\}$ and $\{\eta(t)|t \in T\}$ are identically distributed, since Cov $(\xi(t), \xi(s))$ may fail to equal Cov $(\eta(t), \eta(s))$. For example, let $\eta(t) = e^{i\theta}\,\xi(t)$ where θ is a real number. Let $\mathscr{E}\xi(t) = 0$. then

$$\mathscr{E}\eta(t) = \mathscr{E}\xi(t) = 0$$

and

$$\mathscr{E}\eta(t)\bar{\eta}(s) = \mathscr{E}\xi(t)\bar{\xi}(s) \quad .$$

But if $\mathscr{E}\xi(t)\xi(s) \neq 0$ and θ is incongruent to zero modulo π, then

$$\mathscr{E}\eta(t)\eta(s) = e^{2i\theta}\mathscr{E}\xi(t)\xi(s) \neq \mathscr{E}\xi(t)\xi(s) \quad .$$

In the next two theorems we shall examine this condition on the covariance of $\xi(t)$ and $\xi(s)$. Suppose, then, that $R(t, s)$ is a nonnegative definite function defined on $T \times T$. Then we assert that there exists a complex Gaussian process $\{\xi(t) \mid t \in T\}$ such that $R(t, s)$ is its covariance function and such that $\text{Cov}(\xi(t), \xi(s)) = 0$. Formally stated:

Theorem 4.3. Let T be a linear index set. Let $c(t)$ be a complex-valued function of the real variable t with domain T. Let $R(t, s)$, $t, s \in T$, be nonnegative definite. Then there exists a complex Gaussian process $\{\xi(t) \mid t \in T\}$ such that

$$\mathscr{E}\xi(t) = c(t) \quad , \qquad t \in T \quad , \tag{4.1}$$

$$\text{Cov}(\xi(t), \bar{\xi}(s)) = R(t, s) \quad , \qquad t, s \in T \quad , \tag{4.2}$$

and

$$\text{Cov}(\xi(t), \xi(s)) = 0 \quad , \qquad t, s \in T \quad . \tag{4.3}$$

Proof. Let $t_1, \cdots, t_n$ be distinct points in T. Then the $n \times n$ covariance matrix $\boldsymbol{R} = \| R(t_j, t_k) \|$ is nonnegative definite [see (2.5)]. Also, from (2.3), the $2n \times 2n$ matrix

$$\boldsymbol{C} = \tfrac{1}{2}\,\text{Re}\begin{bmatrix} \boldsymbol{R} & i\boldsymbol{R} \\ i\bar{\boldsymbol{R}} & \boldsymbol{R} \end{bmatrix} \tag{4.4}$$

is nonnegative definite. Now let $x(t)$ and $y(t)$ be real random variables such that $\mathscr{E}x(t) = \text{Re}\, c(t)$, $\mathscr{E}y(t) = \text{Im}\, c(t)$, and such that the covariance matrix of the $2n$-dimensional real random vector $\{x(t_1), \cdots, x(t_n), y(t_1), \cdots, y(t_n)\}$ is $\boldsymbol{C}$. Then if we set

$$\xi(t) = x(t) + iy(t) \quad ,$$

we see that (4.1), (4.2), and (4.3) of the theorem are satisfied.

A characterization of complex Gaussian processes which have the property (4.3) is expressed in the following interesting theorem (see [Gr]).

Theorem 4.4 (Grettenberg's Theorem). Let $\{\xi(t) \mid t \in T\}$ be a complex Gaussian process with mean zero. Then a necessary and sufficient condition that

$$\mathscr{E}\xi(t)\,\xi(s) = 0$$

for all $t, s \in T$ is that $\{\xi(t) \mid t \in T\}$ and $\{e^{i\theta}\xi(t) \mid t \in T\}$ be identically distributed for all real θ.

Proof. Suppose $\{\xi(t) \mid t \in T\}$ and $\{\zeta(t) \mid t \in T\}$ where $\zeta(t) = e^{i\theta}\,\xi(t)$ are identically distributed for all real θ. Then we must have

$$\mathscr{E}\xi(t)\,\xi(s) = \mathscr{E}\zeta(t)\,\zeta(s)$$

for all $t, s \in T$. But

$$\mathscr{E}\zeta(t)\,\zeta(s) = e^{2i\theta}\,\mathscr{E}\xi(t)\,\xi(s) \quad ,$$

and if this is to be true for all real θ, we must have $\mathscr{E}\xi(t)\xi(s) = 0$.

Conversely, if $\mathscr{E}\xi(t)\xi(s) = 0$, then

$$\mathscr{E}\zeta(t)\zeta(s) = e^{2i\theta}\mathscr{E}\xi(t)\xi(s) = 0$$

and

$$\mathscr{E}\zeta(t)\bar{\zeta}(s) = \mathscr{E}e^{i\theta}\xi(t)e^{-i\theta}\bar{\xi}(s) = \mathscr{E}\xi(t)\bar{\xi}(s) \quad .$$

Since $\mathscr{E}\zeta(t) = e^{i\theta}\mathscr{E}\xi(t) = 0$ by hypothesis, we conclude that since $\{\xi(t) \mid t \in T\}$ and $\{\zeta(t) \mid t \in T\}$ are Gaussian processes, they are identically distributed.

In the light of the theorems above we shall, unless explicitly stated to the contrary, hereafter consider only those complex Gaussian processes for which (4.3) holds.

5. The Spectral Distribution Function

Let R be the covariance function of a wide-sense stationary stochastic process. Then if R is continuous, it has an integral representation. Precisely:

Theorem 5.1 (Bochner's Theorem). A function R is continuous and nonnegative definite on $(-\infty, \infty)$ if and only if it may be represented in the form

$$R(t) = \int_{-\infty}^{\infty} e^{i2\pi tf}\, dS(f) \tag{5.1}$$

where S is real, bounded, and monotone increasing.

Proof. As in the proofs of Theorems 3.1 and 3.2 of Chapter I we see that if S is real, bounded, and monotone increasing, then R is continuous and nonnegative definite on $(-\infty, \infty)$.

Now suppose R is continuous and nonnegative definite on $(-\infty, \infty)$. Let $z(t)$ be any complex-valued function of t continuous on $[0, \infty)$. Then for any $A > 0$,

$$\sum_{j,k=1}^{n} R(t_j - t_k)\, z(t_j)\, \bar{z}(t_k)\, (t_j - t_{j-1})\, (t_k - t_{k-1}) \geqq 0$$

where $0 = t_0 < t_1 < \cdot \cdot \cdot < t_n = A$ is a partition of $[0, A]$. The foregoing sum is an approximating sum to the Riemann integral

$$\int_0^A \int_0^A R(t-s)\, z(t)\bar{z}(s) dtds \quad , \tag{5.2}$$

and hence (5.2) is nonnegative.

Now let $z(t) = \exp(-i2\pi ft)$ and define h_A as

$$h_A(f) = \frac{1}{A} \int_0^A \int_0^A R(t-s) \exp[-i2\pi f(t-s)]\, dtds$$

$$= \int_{-\infty}^{\infty} \Delta_A(t)\, R(t) \exp(-i\, 2\pi ft)\, dt \geqq 0$$

where

$$\Delta_A(t) = \begin{cases} 1 - \dfrac{|t|}{A} \ , & |t| \leqq A \ , \\ 0 \ , & |t| > A \ , \end{cases}$$

is a triangular function. Since $\Delta_A(t)R(t)$ is continuous, bounded, and absolutely integrable, and $h_A(f) \geqq 0$ for all $A > 0$, the Fourier inversion theorem implies that

$$\Delta_A(t)\, R(t) = \int_{-\infty}^{\infty} h_A(f)\, e^{i2\pi tf}\, df$$

for all t.

As A increases without limit, $\Delta_A(t)\, R(t)$ approaches $R(t)$ for every t. Since $R(t)$ is continuous by hypothesis, it is certainly continuous at $t = 0$. Now h_A may be interpreted as a (nonnormalized) density function and $\Delta_A R$ as its (nonnormalized) characteristic function. But (see Theorem 3.3 of Chapter I) if a sequence of characteristic functions converges to a limit which is continuous at $t = 0$, the limit is identical with the characteristic function of the limiting distribution function.

Let us interpret Theorem 5.1 in terms of stochastic processes. Suppose, then, $\{\xi(t) | -\infty < t < \infty\}$ is a wide-sense stationary complex process. Let the covariance function R of the process be continuous. Then by Bochner's theorem there exists a real, bounded, monotone increasing function S such that

$$R(t) = \int_{-\infty}^{\infty} e^{i2\pi tf}\, dS(f) \quad .$$

We call S the *spectral distribution function* of the process. Since S is defined up to an additive constant, we may always assume $S(-\infty) = 0$ and $S(\infty) = R(0)$. We shall do so. Without loss of generality we shall also always assume that S is right continuous. (See Section 3 of Chapter I.) If S is absolutely continuous, then $S' = W$ exists almost everywhere (Lebesgue measure) and we call W the *spectral density* of the $\{\xi(t) | -\infty < t < \infty\}$ process. By the Fourier inversion formula

$$W(f) = \int_{-\infty}^{\infty} R(t)\, e^{-i2\pi ft}\, dt \tag{5.3}$$

at every continuity point where W has a right- and left-hand derivative.

If $\{\xi(t) | -\infty < t < \infty\}$ is a *real* process, then R is a *real* function and we may write

$$R(t) = \int_{-\infty}^{\infty} \cos 2\pi tf\, dS(f) \quad .$$

Now define $G(0) = 0$ and

$$G(f) = 2S(f) - R(0) \quad , \qquad f > 0 \quad .$$

Then

$$G(\infty) = S(\infty) = R(0)$$

and

$$R(t) = \int_{0}^{\infty} \cos 2\pi tf\, dG(f) \quad .$$

It is possible to give a heuristic interpretation of Theorem 5.1. We recall that in the theory of random noise the correlation (covariance) function and the power spectrum (spectral density) are represented as a Fourier transform pair (Wiener–Khintchine relations). If we pass a time signal with a "white" spectrum through a linear time-invariant filter, it will shape the spectrum. In fact, the power spectrum of the output signal will be proportional to the square of the modulus of the transfer function of the filter (and nonnegative). By suitably choosing the filter one may approximate almost any nonnegative function. Now the inverse transform of this function is a covariance function, and hence positive definite. Essentially this yields the conclusion of Bochner's theorem.

A useful result in future work is the following lemma.

Lemma 5.1. Let S be a spectral distribution function. Let

$$\int_{-\infty}^{\infty} |f|^{\alpha}\, dS(f) < \infty$$

for some $\alpha > 0$. Then for $\mu > 0$,

$$\int_{-\infty}^{\infty} |\sin \pi hf|^{\mu} dS(f) = 0(|h|^{\gamma}) \tag{5.4}$$

as h approaches zero where $\gamma = \min(\mu, \alpha)$.

Proof. For $h \neq 0$

$$\int_{-\infty}^{\infty} |\sin \pi hf|^{\mu} dS(f)$$
$$= \left(\int_{-\infty}^{-|h|^{-1}} + \int_{-|h|^{-1}}^{|h|^{-1}} + \int_{|h|^{-1}}^{\infty} \right) |\sin \pi hf|^{\mu} dS(f) \quad . \tag{5.5}$$

Now $|\sin x| \leqq |x|$ for all real x. Thus

$$H \equiv \int_{-|h|^{-1}}^{|h|^{-1}} |\sin \pi hf|^{\mu} dS(f) \leqq |\pi h|^{\mu} \int_{-|h|^{-1}}^{|h|^{-1}} |f|^{\mu} dS(f)$$
$$= \pi^{\mu} |h|^{\mu} \int_{-|h|^{-1}}^{|h|^{-1}} |f|^{\gamma} |f|^{\mu-\gamma} \, dS(f)$$

where $\gamma = \min(\mu, \alpha)$. Since $\mu - \gamma \geqq 0$,

$$H \leqq \pi^{\mu} |h|^{\mu} |h|^{\gamma-\mu} \int_{-|h|^{-1}}^{|h|^{-1}} |f|^{\gamma} dS(f) = \pi^{\mu} |h|^{\gamma} \int_{-|h|^{-1}}^{|h|^{-1}} |f|^{\gamma} dS(f) \quad .$$

By hypothesis there exists a constant K, independent of h, such that

$$\int_{-|h|^{-1}}^{|h|^{-1}} |f|^{\gamma} dS(f) < K \quad .$$

Hence $H \leqq \pi^{\mu} K |h|^{\gamma}$ and

$$\int_{-|h|^{-1}}^{|h|^{-1}} |\sin \pi hf|^{\mu} dS(f) = 0(|h|^{\gamma}) \quad . \tag{5.6}$$

Also since

$$|\sin \pi hf|^{\mu} \leqq (hf)^{\nu}$$

for $hf \geqq 1$ and any $\nu \geqq 0$, we have

$$\int_{-\infty}^{-|h|^{-1}} |\sin \pi hf|^{\mu} \, dS(f) \leqq |h|^{\gamma} \int_{-\infty}^{-|h|^{-1}} |f|^{\gamma} dS(f) < C|h|^{\gamma} = 0(|h|^{\gamma}) \tag{5.7}$$

for some constant C independent of h. Similarly

$$\int_{|h|^{-1}}^{\infty} |\sin \pi hf|^{\mu} dS(f) = 0(|h|^{\gamma}) \quad . \tag{5.8}$$

Comparing (5.6), (5.7), (5.8) with (5.5), we see the truth of (5.4).

An important consequence of this lemma is

Corollary 5.1. Let R be a continuous covariance function of a wide-sense stationary stochastic process defined on $(-\infty, \infty)$, and let S be its spectral distribution function. Then if

$$\int_{-\infty}^{\infty} |f|^{\alpha} dS(f) < \infty \tag{5.9}$$

for some $\alpha > 0$, we have

$$2R(0) - R(h) - R(-h) = 0(|h|^{\min(2,\alpha)})$$

and

$$6R(0) - 4R(h) - 4R(-h) + R(2h) + R(-2h) = 0(|h|^{\min(4,\alpha)})$$

as h approaches zero.

Proof. By Bochner's theorem

$$2R(0) - R(h) - R(-h) = 4\int_{-\infty}^{\infty} \sin^2 \pi h f \, dS(f)$$

and

$$6R(0) - 4R(h) - 4R(-h) + R(2h) + R(-2h)$$

$$= 16\int_{-\infty}^{\infty} \sin^4 \pi h f \, dS(f) \quad .$$

Now apply Lemma 5.1

Let R be a covariance function continuous on the infinite interval $(-\infty, \infty)$. Then by Theorem 5.1

$$R(t) = \int_{-\infty}^{\infty} e^{i2\pi t f} \, dS(f)$$

where S is real, bounded, and monotone increasing. As usual we shall assume $S(-\infty) = 0$ and $S(\infty) = R(0)$. What we would like to do now is to show the relationship between the moments of the spectrum and the derivatives of the covariance function. Precisely we shall prove:

Theorem 5.2. Let R be a covariance function continuous on the infinite interval $(-\infty, \infty)$. Let S be the corresponding spectral distribution function. Then if the integral

$$\int_{-\infty}^{\infty} |f|^n dS(f) \quad , \qquad n = 0, 1, 2, \cdots \tag{5.10}$$

converges, we have

(i) $R^{(n)}(t)$ exists on $(-\infty, \infty)$

$$\text{(ii)} \quad R^{(n)}(t) = (i2\pi)^n \int_{-\infty}^{\infty} f^n e^{i2\pi tf} dS(f) \tag{5.11}$$

(iii) $R^{(n)}(t)$ is uniformly continuous on $(-\infty, \infty)$.

Proof. The proof is by induction. By Bochner's theorem, (i) and (ii) are true for $n = 0$. Now suppose (5.11) is true for $n = k$ and that

$$\int_{-\infty}^{\infty} |f|^{k+1} dS(f) < \infty \quad .$$

Then if we can show that given any $\varepsilon > 0$ there exists a $\delta > 0$ such that

$$\lambda_h = \left| \frac{R^{(k)}(t+h) - R^{(k)}(t)}{h} - (i2\pi)^{k+1} \int_{-\infty}^{\infty} e^{i2\pi tf} f^{k+1}\, dS(f) \right|$$

is less than ε for all h such that $0 < |h| < \delta$, then we shall have established (i) and (ii) for $n = k + 1$.

Since (5.11) holds for $n = k$ we may write

$$\lambda_h = (2\pi)^{k+1} \left| \int_{-\infty}^{\infty} e^{i2\pi tf} \left[\left(\frac{\sin \pi hf}{\pi hf} \right) e^{i\pi hf} - 1 \right] f^{k+1}\, dS(f) \right| \quad ,$$

and for any $A > 0$,

$$\lambda_h \leqq (2\pi)^{k+1} \left[2 \int_{-\infty}^{-A} |f|^{k+1} dS(f) + 2 \int_{A}^{\infty} |f|^{k+1} dS(f) \right.$$
$$\left. + \int_{-A}^{A} \left| \left(\frac{\sin \pi hf}{\pi hf} \right) e^{i\pi hf} - 1 \right| |f|^{k+1} dS(f) \right] \quad .$$

Thus if we choose $A > 0$ sufficiently large, and then choose $h \neq 0$ sufficiently small in absolute value, we can make λ_h less than any preassigned positive number.

The uniform continuity of $R^{(n)}(t)$, $n = 0, 1, 2, \cdots$, is proved as in Theorem 3.1 of Chapter I.

Corollary 5.2. If

$$\int_{-\infty}^{\infty} |f|^n dS(f) < \infty \quad ,$$

then

$$R^{(n)}(0) = (i2\pi)^n \int_{-\infty}^{\infty} f^n dS(df) \quad .$$

One may also show that (see [CL, page 134], [Cr, page 90]):

Theorem 5.3. Let R be a covariance function continuous on $(-\infty, \infty)$. Let S be the corresponding spectral distribution function. Then

$$\int_{-\infty}^{\infty} f^{2k} dS(f) < \infty$$

if and only if $R^{(2k)}(0)$ exists.

6. Continuity of Sample Functions

Let $\{x(t) | t \in T\}$ be a real stochastic process defined on a probability space $\mathscr{P} = (\Omega, \mathscr{F}, P)$. We would like to know under what conditions $x(t)$ is a continuous function of t on T for almost all ω in Ω. The fundamental result is given in Theorem 6.1. It yields sufficient conditions which guarantee the existence of an equivalent stochastic process whose sample functions are continuous with probability one.

Theorem 6.1. Let $\{x(t) | t \in [0, 1]\}$ be a real stochastic process. Suppose that for all t and $t + h$ in $[0, 1]$ we have

$$P[|x(t + h) - x(t)| \geqq g(h)] \leqq q(h)$$

where $g(h)$ and $q(h)$ are nonnegative monotone decreasing even functions of h as h approaches zero such that $\sum_{n=1}^{\infty} g(2^{-n}) < \infty$ and

$$\sum_{n=1}^{\infty} 2^n q(2^{-n}) < \infty \quad .$$

Then there exists an equivalent stochastic process $\{x^*(t) | t \in [0, 1]\}$ whose sample functions are continuous with probability one on $[0, 1]$.

Proof. For every $n \geqq 1$ let π_n be a partition of $[0, 1]$ into 2^n equal subintervals. That is

$$0, h, 2h, \cdot\cdot\cdot, 2^n h = 1$$

are the points of subdivision. Define the continuous piecewise linear function $x_n(t)$ as

$$x_n(t) = x(k2^{-n}) + 2^n[x((k + 1)2^{-n}) - x(k2^{-n})] (t - k2^{-n})$$

for

$$k2^{-n} \leqq t \leqq (k + 1)2^{-n} \quad , \qquad 0 \leqq k \leqq 2^n - 1 \quad .$$

Then $x_n(t)$ is a polygonal approximation to $x(t)$ on [0, 1].

Also

$$x_{n+1}(t) = x(k2^{-n}) + 2^{n+1}[x((k + \tfrac{1}{2})2^{-n}) - x(k2^{-n})]\,(t - k2^{-n})$$

for $t \in I_1$ where $I_1 = [k2^{-n}, (k + \frac{1}{2})2^{-n}]$ and

$$x_{n+1}(t) = x((k + \tfrac{1}{2})2^{-n}) + 2^{n+1}[x((k + 1)2^{-n}) - x((k + \tfrac{1}{2})2^{-n})]\,(t - (k + \tfrac{1}{2})2^{-n})$$

for $t \in I_2$ where $I_2 = [(k + \frac{1}{2})2^{-n}, (k + 1)2^{-n}]$. Therefore, since

$$|x_{n+1}(t) - x_n(t)| \leqq \tfrac{1}{2}|x(k2^{-n}) - 2x((k + \tfrac{1}{2})2^{-n}) + x((k + 1)2^{-n})|$$

for $t \in I_1$ and also for $t \in I_2$, we certainly have

$$|x_{n+1}(t) - x_n(t)| \leqq |x(k2^{-n}) - x((k + \tfrac{1}{2})2^{-n})| + |x((k + \tfrac{1}{2})2^{-n}) - x((k + 1)2^{-n})|$$

for $t \in I_1 \cup I_2 = I$. Thus

$$\begin{aligned} P[\max_{t\in I}|x_{n+1}(t) - x_n(t)| \geqq g(2^{-n-1})] &\leqq P[|x(k2^{-n}) - x((k + \tfrac{1}{2})2^{-n})| \geqq g(2^{-n-1})] \\ &\quad + P[|x((k + \tfrac{1}{2})2^{-n}) - x((k + 1)2^{-n})| \geqq g(2^{-n-1})] \leqq 2q(2^{-n-1}) \end{aligned}$$

—the last inequality by hypothesis. (Note that we have used the fact that g is an even function.)

Now there are 2^n subintervals of [0, 1]. Therefore

$$P[\max_{t\in[0,1]}|x_{n+1}(t) - x_n(t)| \geqq g(2^{-n-1})] \leqq 2^{n+1}q(2^{-n-1}) \quad .$$

But by hypothesis $\sum_{n=1}^{\infty} 2^n q(2^{-n})$ converges. Therefore, by Lemma 4.1 of Chapter I, there exists a random variable, say $x^*(t)$, such that $\{x_n(t)\}$ converges with probability one to $x^*(t)$. Furthermore, the convergence is uniform. And since each $x_n(t)$ is continuous, $x^*(t)$ is continuous with probability one.

For $t = k2^{-n}$, $0 \leqq k \leqq 2^n$, $n = 1, 2, \cdots$, we have $x_n(t) = x(t)$, and thus $x^*(k2^{-n}) = x(k2^{-n})$ with probability one. For $t \neq k2^{-n}$ for any k and n we have $t \in (k2^{-n}, (k + 1)2^{-n})$. Now for every n there is a k_n such that $0 < t - k_n2^{-n} < 2^{-n}$ and

$$t = \lim_{n\to\infty} k_n 2^{-n} \quad .$$

Thus

$$P[|x(k_n2^{-n}) - x(t)| \geqq g(t - k_n2^{-n})] \leqq q(t - k_n2^{-n}) \leqq q(2^{-n})$$

since q is monotone. By Theorem 2.4 of Chapter I (Borel–Cantelli lemma) $x(k_n 2^{-n})$ converges with probability one to $x(t)$.

But x^* is continuous with probability one. Thus $x^*(k_n 2^{-n})$ converges with probability one to $x^*(t)$. Since $x(k_n 2^{-n}) = x^*(k_n 2^{-n})$, we see that $x^*(k_n 2^{-n})$ converges with probability one to $x(t)$. Thus $x(t) = x^*(t)$ with probability one for all $t \in [0, 1]$.

Clearly the restriction of the parameter set to the unit interval is unessential. If $\{\xi(t) | t \in T\}$ is a complex process, then we may write $\xi(t) = x(t) + iy(t)$ and apply Theorem 6.1 to $x(t)$ and $y(t)$, respectively.

Theorem 6.1 is not a particularly convenient criterion to apply. However, in the cases where the process is stationary, sufficient conditions for continuity can be very easily and simply stated in terms of the covariance function. More general theorems involving the covariance function, for both stationary and nonstationary processes, exist. Theorems 6.2 and 6.3 transfer the conditions for continuity to the covariance function in the cases in which the process is weakly stationary.

Theorem 6.2. Let $\{x(t) | t \in T\}$ be a real wide-sense stationary stochastic process with covariance function R. If for some $\alpha > 0$,

$$R(h) = R(0) - 0(|h|^{1+\alpha}) \tag{6.1}$$

as h approaches zero, then there exists an equivalent process which possesses continuous sample functions with probability one.

Proof. In the notation of Theorem 6.1 let

$$g(h) = |h|^{\beta} \quad , \qquad q(h) = |h|^{1+\alpha-2\beta} \tag{6.2}$$

where $0 < \beta < \frac{1}{2}\alpha$. Then $\sum_{n=1}^{\infty} g(2^{-n})$ converges monotonely and

$$\sum_{n=1}^{\infty} 2^n q(2^{-n}) = \sum_{n=1}^{\infty} 2^{-n(\alpha-2\beta)} < \infty \quad . \tag{6.3}$$

Now by the Tschebyscheff inequality

$$P[|x(t+h) - x(t)| \geqq g(h)] \leqq \frac{1}{g^2(h)} \mathscr{E}[x(t+h) - x(t)]^2 \tag{6.4}$$

for all t, $t + h$ in T. But

$$\mathscr{E}[x(t+h) - x(t)]^2 = 2[R(0) - R(h)] \quad .$$

Since $R(h) = R(0) - 0(|h|^{1+\alpha})$ by hypothesis, there exists a constant K independent of h such that

$$R(0) - R(h) < K|h|^{1+\alpha} \quad .$$

Thus from (6.4)

$$P[|x(t+h) - x(t)| \geqq |h|^{\beta}] < 2K|h|^{1+\alpha-2\beta} \quad .$$

From this inequality, the fact that $\sum_{n=1}^{\infty} 2^{-n\beta}$ converges, and (6.3) we see that the conditions of Theorem 6.1 are met.

If we make the additional assumption that the $x(t)$ process be *Gaussian,* then we may weaken the hypothesis [see (6.1)] to

$$R(h) = R(0) - 0(|h|^{\alpha}) \quad .$$

That is, we can replace $|h|^{1+\alpha}$ by $|h|^{\alpha}$.

Theorem 6.3. Let $\{x(t) | -\infty < t < \infty\}$ be a real stationary Gaussian process with covariance function R. If for some $\alpha > 0$,

$$R(h) = R(0) - 0(|h|^{\alpha})$$

as h approaches zero, then there exists an equivalent process which possesses continuous sample functions with probability one.

Proof. In the notation of Theorem 6.1 let $g(h) = |h|^{\beta}$ where $0 < \beta < \frac{1}{2}\alpha$. Then $\sum_{n=1}^{\infty} g(2^{-n})$ converges monotonely.

Now

$$P[|x(t+h) - x(t)| \geqq |h|^{\beta}] = 1 - 2 \operatorname{erf} \frac{|h|^{\beta}}{\sigma}$$

where

$$\operatorname{erf} v = \frac{1}{\sqrt{2\pi}} \int_0^v \exp(-\tfrac{1}{2}u^2)\, du$$

is the error function and

$$\begin{aligned} \sigma^2 &= \operatorname{Var}[x(t+h) - x(t)] = \mathscr{E}[x(t+h) - x(t)]^2 \\ &= 2[R(0) - R(h)] \quad . \end{aligned}$$

For any $h \neq 0$ there exists a constant K independent of h such that

$$R(0) - R(h) < K|h|^{\alpha} \quad .$$

Hence

$$P[|x(t+h) - x(t)| \geqq |h|^{\beta}] < 1 - 2 \operatorname{erf} (2K)^{-\frac{1}{2}}|h|^{\beta-(\alpha/2)} \quad .$$

But from the asymptotic formula for the error function we have

$$1 - 2 \operatorname{erf} v < \left(\frac{1}{v}\right) \exp\left(-\tfrac{1}{2}v^2\right)$$

for all $v > 0$. Thus

$$P[|x(t+h) - x(t)| \geqq |h|^{\beta}] < \sqrt{2K}\, |h|^{(\alpha/2)-\beta} \exp\left(\frac{-|h|^{2\beta-\alpha}}{4K}\right) \quad .$$

Since

$$\sum_{n=1}^{\infty} 2^{n\gamma} \exp\left(-2^{n\mu}\right)$$

converges for any γ and any positive μ, Theorem 6.1 may be invoked to complete the proof.

Suppose $\{\xi(t) | -\infty < t < \infty\}$ is a *complex* stationary Gaussian process,

$$\begin{aligned} \operatorname{Cov}(\xi(t), \bar{\xi}(s)) &= R(t-s) \quad , \\ \operatorname{Cov}(\xi(t), \xi(s)) &= 0 \end{aligned} \tag{6.5}$$

(see Section 4). If we write

$$\xi(t) = x(t) + iy(t) \quad , \tag{6.6}$$

then the continuity of $x(t)$ and that of $y(t)$ imply the continuity of $\xi(t)$. Thus if we can show that corresponding to the real processes $\{x(t)\}$ and $\{y(t)\}$ there exist equivalent processes which possess continuous sample functions with probability one, then we shall have proved the existence of a process equivalent to $\{\xi(t)\}$ which possesses continuous sample functions with probability one.

From (6.5) and (6.6) we conclude that

$$\operatorname{Cov}(x(t), x(s)) = \tfrac{1}{2} \operatorname{Re} R(t-s) = \operatorname{Cov}(y(t), y(s)) \tag{6.7}$$

and

$$\operatorname{Cov}(x(t), y(s)) = -\tfrac{1}{2} \operatorname{Im} R(t-s) = -\operatorname{Cov}(x(s), y(t)) \quad . \tag{6.8}$$

Now suppose

$$R(h) + R(-h) = 2R(0) - 0(|h|^{\alpha}) \quad , \qquad \alpha > 0 \quad ,$$

as h approaches zero. Then

$$C(h) = C(0) - 0(|h|^{\alpha})$$

where

$$C(h) = \tfrac{1}{2}\ \mathrm{Re}\ R(h) = \mathrm{Cov}\,(x(t+h), x(t)) = \mathrm{Cov}\,(y(t+h), y(t)) \quad .$$

Hence we have the following corollary to Theorem 6.3.

Corollary 6.1. Let $\{\xi(t) \mid -\infty < t < \infty\}$ be a complex stationary Gaussian process with covariance function R. If for some $\alpha > 0$,

$$R(h) + R(-h) = 2R(0) - 0(|h|^{\alpha}) \tag{6.9}$$

as h approaches zero, then there exists an equivalent complex process which possesses continuous sample functions with probability one.

If R is a covariance function continuous on $(-\infty, \infty)$, then by Bochner's theorem its spectral distribution function exists. By Corollary 5.1 we can transfer the conditions in Theorems 6.2 and 6.3 and Corollary 6.1 on the covariance function to the spectral distribution function. This leads to:

Theorem 6.4. Let $\{x(t) \mid -\infty < t < \infty\}$ be a real wide-sense stationary stochastic process with continuous covariance function R and spectral distribution function S. If for some $\beta > 0$

$$\int_{-\infty}^{\infty} |f|^{1+\beta}\, dS(f) < \infty \quad , \tag{6.10}$$

then there exists an equivalent process which possesses continuous sample functions with probability one.

Proof. Let

$$\gamma = \min\,(1, \beta) \quad . \tag{6.11}$$

Then from (6.10),

$$\int_{-\infty}^{\infty} |f|^{1+\gamma} dS(f) < \infty$$

and from Corollary 5.1

$$R(0) - R(h) = 0(|h|^{\min(2,1+\gamma)})$$

as h approaches zero. But

$$\min(2, 1+\gamma) = 1 + \gamma \quad .$$

Thus

$$R(h) = R(0) - 0(|h|^{1+\gamma})$$

as h approaches zero. Now apply Theorem 6.2.

Similarly for Gaussian processes we have:

Theorem 6.5. Let $\{\xi(t) | -\infty < t < \infty\}$ be a complex stationary Gaussian process with continuous covariance function R and spectral distribution function S. If for some $\beta > 0$

$$\int_{-\infty}^{\infty} |f|^{\beta} \, dS(f) < \infty \tag{6.12}$$

then there exists an equivalent process which possesses continuous sample functions with probability one.

Proof. Let

$$\gamma = \min (1, \beta) \quad .$$

Then from (6.12)

$$\int_{-\infty}^{\infty} |f|^{\gamma} dS(f) < \infty$$

and from Corollary 5.1

$$2R(0) - R(h) - R(-h) = 0(|h|^{\gamma})$$

as h approaches zero. Corollary 6.1 completes the proof.

In general, if $\{\xi(t) | t \in T\}$ is an arbitrary complex process whose real and imaginary parts are wide-sense stationary, then we may apply the theorems above to determine whether $\{\xi(t) | t \in T\}$ has continuous sample functions.

Suppose $\{\xi(t) | -\infty < t < \infty\}$ is a mean zero wide-sense stationary complex stochastic process with covariance function R and

$$R(h) = R(0) - 0(|h|^{\gamma}) \quad . \tag{6.13}$$

Then (6.13) implies

$$R(h) + R(-h) = 2R(0) - 0(|h|^{\gamma}) \tag{6.14}$$

but not conversely. Thus (6.13) implies

$$\text{Re } R(h) = R(0) - 0(|h|^{\gamma}) \tag{6.15}$$

and

$$\text{Im } R(h) = -0(|h|^{\gamma}) \tag{6.16}$$

while (6.14) implies only (6.15) and not (6.16).

If we write

$$\xi(t) = x(t) + iy(t) \quad ,$$

then

$$\text{Re } R(t - s) = \mathscr{E}[x(t)x(s) + y(t)y(s)]$$

and

$$\text{Im } R(t - s) = \mathscr{E}[y(t)x(s) - y(s)x(t)] \quad .$$

Thus the properties of Re R do not imply anything about $\mathscr{E}x(t)x(s)$ and $\mathscr{E}y(t)y(s)$ individually. For instance, if $\gamma > 1$, then certainly Re R is continuous; however, we may not infer that $\mathscr{E}x(t)x(s)$ or $\mathscr{E}y(t)y(s)$ is also continuous.

7. Mean Square Continuity

Let $\{\xi(t) | t \in T\}$ be a complex stochastic process defined on a probability space $\mathscr{P}$. We shall assume, in this and the next section, that T is an interval (finite or infinite). If $\{\xi(t)\}$ converges in mean square to $\xi(t_0)$ as t approaches $t_0 \in T$, then we shall say that $\{\xi(t) | t \in T\}$ is *continuous in mean square* at $t = t_0$. We may write

$$\lim_{h \to 0} \mathscr{E} | \xi(t_0 + h) - \xi(t_0) |^2 = 0$$

or

$$\underset{t \to t_0}{\text{l.i.m.}}\ \xi(t) = \xi(t_0)$$

to express this fact. If $\{\xi(t)\}$ is continuous in mean square for all $t \in I \subset T$, then we shall say $\{\xi(t)\}$ is *continuous in mean square on I*. The continuity of the covariance function of the $\{\xi(t) | t \in T\}$ process may be used to establish mean square continuity.

Theorem 7.1. Let $R(t, s)$ be the covariance function of the mean zero stochastic process $\{\xi(t) | t \in T\}$. Then $\{\xi(t)\}$ is continuous in mean square at $t = t_0 \in T$ if and only if $R(t, s)$ is continuous at the point $(t_0, t_0) \in T \times T$.

Proof. Suppose $R(t, s)$ is continuous at $t = s = t_0 \in T$. Then the identity

$$\mathscr{E} | \xi(t + h) - \xi(t) |^2 = R(t + h, t + h) - R(t, t + h) - R(t + h, t) + R(t, t) \quad (7.1)$$

implies

$$\underset{h \to 0}{\text{l.i.m.}}\ \xi(t_0 + h) = \xi(t_0) \quad .$$

Conversely, suppose $\{\xi(t)\}$ is continuous in mean square at $t = t_0$. Then the Schwarz inequality applied to both terms on the right-hand side of the identity

$$R(t + h, t + k) - R(t, t) = \mathscr{E}[\xi(t + h) - \xi(t)]\bar{\xi}(t + k) + \mathscr{E}\xi(t)\,[\bar{\xi}(t + k) - \bar{\xi}(t)]$$

implies

$$\lim_{h,k \to 0} R(t_0 + h, t_0 + k) = R(t_0, t_0) \quad .$$

If R is a covariance function, then

$$|R(t, s)|^2 \leqq R(t, t)\, R(s, s) \quad .$$

Hence it is not surprising that the continuity properties of $R(t, s)$ for $t \neq s$ may be deduced from the continuity of $R(t, s)$ on the diagonal $t = s$. In particular we have the following theorem.

Theorem 7.2. Let R be the covariance function of the mean zero stochastic process $\{\xi(t) | t \in T\}$. Then if $R(t, s)$ is continuous at every diagonal point $t = s \in T$, the covariance function R is continuous on $T \times T$.

Proof. By the previous theorem and Theorem 6.3 of Chapter I, if $R(t, s)$ is continuous at (t, t) and (s, s), then we may conclude that

$$\mathscr{E}\xi(t + h)\, \bar{\xi}(s + k) = R(t + h, s + k)$$

converges to $\mathscr{E}\xi(t)\bar{\xi}(s) = R(t, s)$ as h and k independently approach zero.

Corollary 7.1. Let R be the covariance function of the wide-sense stationary process $\{\xi(t) | t \in T\}$. Then $\{\xi(t)\}$ is continuous in mean square for all $t \in T$ if and only if $R(t)$ is continuous at $t = 0$. Furthermore, if $R(t)$ is continuous at $t = 0$, then $R(t)$ is continuous for all $t \in \text{Dom } R$.

Proof. By the continuity of $R(t)$ at $t = 0$

$$\mathscr{E}|\xi(t + h) - \xi(t)|^2 = 2R(0) - R(h) - R(-h) \to 0$$

as $h \to 0$. Thus $\{\xi(t)\}$ is continuous in mean square for all $t \in T$. Hence, by Theorem 6.3 of Chapter I,

$$\lim_{h \to 0} \mathscr{E}\xi(t + s + h)\, \bar{\xi}(s) = \mathscr{E}\xi(t + s)\, \bar{\xi}(s) = R(t) + |\mu|^2$$

where $\mu = \mathscr{E}\xi(t)$. But $\mathscr{E}\xi(t + s + h)\, \bar{\xi}(s) = R(t + h) + |\mu|^2$. Hence $R(t)$ is continuous for all $t \in \text{Dom } R$.

The converse is as in Theorem 7.1.

If R is a real function with the property that

$$R(h) = R(0) - 0(|h|^{\beta})$$

as h approaches zero for some $\beta > 0$, then we see that $R(h)$ is continuous at $h = 0$. Using this fact together with Corollary 7.1 and Theorems 5.1, 5.2, and 6.2, we have:

Theorem 7.3. Let $\{x(t)| - \infty < t < \infty\}$ be a real wide-sense stationary process with covariance function R. If for some $\alpha > 0$

$$R(h) = R(0) - 0(|h|^{1+\alpha}) \tag{7.2}$$

as h approaches zero, then

(i) R is uniformly continuous on $(-\infty, \infty)$
(ii) $\{x(t)\}$ is continuous in mean square on $(-\infty, \infty)$
(iii) there exists an equivalent process which possesses continuous sample functions with probability one
(iv) the spectral distribution function S of $\{x(t)\}$ exists and

$$R(t) = \int_{-\infty}^{\infty} e^{i2\pi tf} dS(f) \quad .$$

If $R = C + iD$ is a covariance function, then $|R(h)| \leqq R(0)$, $C(h) = C(-h)$, and $D(h) = -D(-h)$. Thus $D(0) = 0$ and $C(0) = R(0)$. Now suppose $C(h)$ is continuous at $h = 0$. Then

$$[C^2(h) + D^2(h)]^{1/2} = |R(h)| \leqq R(0) = C(0) \tag{7.3}$$

implies that $D(h)$ is continuous at $h = 0$. But if $C(h)$ and $D(h)$ are continuous at $h = 0$, then $R(h)$ is continuous at $h = 0$.

Now suppose

$$R(h) + R(-h) = 2R(0) - 0(|h|^{\alpha}) \tag{7.4}$$

as h approaches zero for some $\alpha > 0$. Then Re $R(h)$ is continuous at $h = 0$, and hence by the argument above, $R(h)$ is continuous at $h = 0$. Thus we see that if $\{\xi(t)| -\infty < t < \infty\}$ is a complex stationary Gaussian process with covariance function R, and if (7.4) holds, then by Corollaries 6.1, 7.1 and Theorems 5.1, 5.2 the conclusions (i)–(iv) of Theorem 7.3 hold (with x replaced by ξ). Formally stated:

Corollary 7.2. Let $\{\xi(t)| - \infty < t < \infty\}$ be a complex stationary Gaussian process with covariance function R. If for some $\alpha > 0$

$$R(h) + R(-h) = 2R(0) - 0(|h|^{\alpha})$$

as h approaches zero, then the conclusions of Theorem 7.3 hold (with x replaced by ξ).

By Bochner's theorem, if the spectral distribution function S of a wide-sense stationary process exists, then the covariance function R of the process is continuous. Hence by Corollary 7.1 the process is mean square continuous. Thus we have

Corollary 7.3. Let the spectral distribution function S of the wide-sense stationary process $\{\xi(t) | -\infty < t < \infty\}$ exist. Then $\{\xi(t)\}$ is continuous in mean square on $(-\infty, \infty)$.

Note that the existence of S is tantamount to the statement

$$\int_{-\infty}^{\infty} dS(f) < \infty \quad .$$

Let $\{x(t) | t \in T\}$ be a stochastic process. Suppose there exists a process $\{x^*(t) | t \in T\}$ equivalent to $\{x(t)\}$ which possesses continuous sample functions with probability one. Then $x^*(t_0) = x(t_0)$ with probability one for all $t_0 \in T$. Let $\{h_n\}$ be a sequence of real numbers converging to zero and let

$$\lambda_n = x^*(t_0 + h_n) \quad , \qquad t_0,\, t_0 + h_n \in T \quad .$$

Then

$$\lim_{n \to \infty} \lambda_n = x^*(t_0) = [x(t_0)]_{\mathrm{ae}} \quad .$$

Now suppose $\{x(t) | t \in T\}$ is also continuous in mean square. Let

$$\mu_n = x(t_0 + h_n) \quad , \qquad t_0,\, t_0 + h_n \in T \quad .$$

Then

$$\underset{n \to \infty}{\text{l.i.m.}}\ \mu_n = [x(t_0)]_{\mathrm{ms}}$$

and by Corollary 7.1 of Chapter I,

$$[x(t_0)]_{\mathrm{ms}} = [x(t_0)]_{\mathrm{ae}}$$

with probability one.

8. Differentiability of Random Variables

The basic theorem regarding the existence of derivatives of sample functions is given by (see [CL, page 67])

Theorem 8.1. Let $\{x(t)\,|\,t \in T\}$ be a real stochastic process. Suppose that for all t, $t+h$, $t-h$ in T we have

$$P[\,|2x(t) - x(t+h) - x(t-h)| \geqq g(h)] \leqq q(h)$$

where $g(h)$ and $q(h)$ are nonnegative monotone decreasing even functions of h as h approaches zero such that $\sum_{n=1}^{\infty} 2^n g(2^{-n}) < \infty$ and

$$\sum_{n=1}^{\infty} 2^n q(2^{-n}) < \infty \quad .$$

Then there exists an equivalent stochastic process $\{x^*(t)\,|\,t \in T\}$ whose sample functions have continuous derivatives with probability one on T.

Conditions on the covariance function of a stationary stochastic process sufficient to guarantee the existence of differentiable sample functions are given in the next two theorems.

Theorem 8.2. Let $\{x(t)\,|\,t \in T\}$ be a real wide-sense stationary stochastic process with covariance function R. If for some $\alpha > 0$,

$$R(h) = R(0) - \lambda h^2 + 0(|h|^{3+\alpha}) \quad , \qquad \lambda > 0 \quad ,$$

as h approaches zero, then there exists an equivalent process which possesses continuous sample derivatives with probability one.

Proof. In the notation of Theorem 8.1 let

$$g(h) = |h|^{\beta} \quad , \qquad q(h) = |h|^{3+\alpha-2\beta} \quad ,$$

where $1 < \beta < 1 + \frac{1}{2}\alpha$. Then $\{g(2^{-n})\}$ and $\{q(2^{-n})\}$ satisfy the convergence requirements of Theorem 8.1. Now by the Tschebyscheff inequality

$$P[\,|2x(t) - x(t+h) - x(t-h)| \geqq g(h)]$$
$$\leqq \frac{1}{g^2(h)} \mathscr{E}[2x(t) - x(t+h) - x(t-h)]^2 \quad .$$

But

$$\mathscr{E}[2x(t) - x(t+h) - x(t-h)]^2 = 6R(0) - 8R(h) + 2R(2h)$$
$$= 0(|h|^{3+\alpha}) \quad .$$

The proof now proceeds as in Theorem 6.2.

If we make the additional assumption that the process be *Gaussian* (cf.

Theorems 6.2 and 6.3), then we can weaken the requirements on the covariance function.

Theorem 8.3. Let $\{x(t) | -\infty < t < \infty\}$ be a real stationary Gaussian process with covariance function R. If for some $\alpha > 0$,

$$R(h) = R(0) - \lambda h^2 + 0(|h|^{2+\alpha}) \quad , \qquad \lambda > 0 \quad ,$$

as h approaches zero, then there exists an equivalent process which possesses continuous sample derivatives with probability one.

Proof. In the notation of Theorem 8.1 let $g(h) = |h|^\beta$ where $1 < \beta < 1 + \frac{1}{2}\alpha$. Then $\sum_{n=1}^{\infty} 2^n g(2^{-n})$ converges. Now

$$P[|2x(t) - x(t+h) - x(t-h)| \geqq |h|^\beta] = 1 - 2 \operatorname{erf} \frac{|h|^\beta}{\sigma}$$

where

$$\begin{aligned} \sigma^2 &= \operatorname{Var}[2x(t) - x(t+h) - x(t-h)] \\ &= 6R(0) - 8R(h) + 2R(2h) = 0(|h|^{2+\alpha}) \quad . \end{aligned}$$

The proof now proceeds as in Theorem 6.3.

If $\{\xi(t) | -\infty < t < \infty\}$ is a complex stationary Gaussian process with covariance function R, and if

$$6R(0) - 4R(h) - 4R(-h) + R(2h) + R(-2h) = 0(|h|^{2+\alpha})$$

for some $\alpha > 0$ as h approaches zero, then

$$3C(0) - 4C(h) + C(2h) = 0(|h|^{2+\alpha})$$

where $C = \frac{1}{2} \operatorname{Re} R$. Hence by Theorem 8.3 the real and imaginary parts of ξ have continuous derivatives a.e. Thus:

Corollary 8.1. Let $\{\xi(t) | -\infty < t < \infty\}$ be a complex stationary Gaussian process with covariance function R. If for some $\alpha > 0$,

$$6R(0) - 4R(h) - 4R(-h) + R(2h) + R(-2h) = 0(|h|^{2+\alpha})$$

as h approaches zero, then there exists an equivalent complex process which possesses continuous sample derivatives with probability one.

We now use Theorem 5.1 to prove theorems analogous to Theorems 6.4 and 6.5 which yield sufficient conditions for the existence of differentiable sample functions.

Theorem 8.4. Let $\{x(t) | -\infty < t < \infty\}$ be a real wide-sense stationary stochastic process with continuous covariance function R and spectral distribution function S. If for some $\beta > 0$

$$\int_{-\infty}^{\infty} |f|^{3+\beta} \, dS(f) < \infty \quad , \tag{8.1}$$

then there exists an equivalent process which possesses continuous sample derivatives with probability one.

Proof. Let

$$\gamma = \min(1, \beta) \quad . \tag{8.2}$$

Then from (8.1)

$$\int_{-\infty}^{\infty} |f|^{3+\gamma} \, dS(f) < \infty \quad .$$

But by Bochner's theorem and Corollary 5.1,

$$6R(0) - 8R(h) + 2R(2h) = 0(|h|^{3+\gamma}) \quad .$$

Now apply Theorem 8.2.

For the Gaussian case we have

Theorem 8.5. Let $\{\xi(t) | -\infty < t < \infty\}$ be a complex stationary Gaussian process with continuous covariance function R and spectral distribution function S. If for some $\beta > 0$

$$\int_{-\infty}^{\infty} |f|^{2+\beta} \, dS(f) < \infty \tag{8.3}$$

then there exists an equivalent process which possesses continuous sample derivatives with probability one.

Proof. Let

$$\gamma = \min(1, \beta) \quad .$$

Then from (8.3)

$$\int_{-\infty}^{\infty} |f|^{2+\gamma} dS(f) < \infty$$

and from Corollary 5.1

$$6R(0) - 4R(h) - 4R(-h) + R(2h) + R(-2h) = 0(|h|^{2+\gamma}) \quad .$$

But if $C = \frac{1}{2}\operatorname{Re} R$, then

$$6C(0) - 8C(h) + 2C(2h) = 0(|h|^{2+\gamma}) \quad .$$

The proof now proceeds as in Theorem 8.3.

We now turn to the consideration of *mean square differentiability*. Let $\{\xi(t) \mid t \in T\}$ be a complex stochastic process with mean zero. If there exists a stochastic process $\{\eta(t) \mid t \in T\}$ such that

$$\frac{\xi(t_0 + h) - \xi(t_0)}{h} \quad , \qquad t_0,\, t_0 + h \in T \quad ,$$

converges in mean square to $\eta(t_0)$, as h approaches zero, then $\eta(t_0)$ is called the *mean square derivative of* $\xi(t)$ *at* $t = t_0$, and we write $\xi'(t_0) = \eta(t_0)$. If $\{\xi(t)\}$ is mean square differentiable for all $t \in I \subset T$, then we shall say $\{\xi(t)\}$ is *differentiable in mean square on I*. Just as for continuity (see Theorems 7.1 and 7.2) the differentiability of the covariance function of the $\{\xi(t) \mid t \in T\}$ process may be used to establish sufficient conditions under which the process has a mean square derivative.

Theorem 8.6. Let $R(t, s)$ be the covariance function of the mean zero stochastic process $\{\xi(t) \mid t \in T\}$. Then $\xi(t)$ has a mean square derivative at $t = t_0 \in T$ provided $\partial^2 R/\partial t \partial s$ exists in a neighborhood of (t_0, t_0) and is continuous at (t_0, t_0).

Proof. Since R is a covariance function, $R(t, s) = \bar{R}(s, t)$. Thus the existence or continuity of $\partial^2 R/\partial t \partial s$ implies the existence or continuity, respectively, of $\partial^2 R/\partial s \partial t$. Now consider the identity

$$\begin{aligned} &\mathscr{E}\left[\frac{\xi(t_0 + h) - \xi(t_0)}{h}\right]\left[\frac{\bar{\xi}(t_0 + k) - \bar{\xi}(t_0)}{k}\right] \\ &= \frac{1}{hk}\{[R(t_0 + h, t_0 + k) - R(t_0, t_0 + k)] \\ &\qquad - [R(t_0 + h, t_0) - R(t_0, t_0)]\} \quad . \end{aligned} \tag{8.4}$$

Since $\partial^2 R/\partial t \partial s$ exists, $\partial R/\partial s$ exists in a neighborhood of (t_0, t_0). Thus by the Law of the mean

$$\begin{aligned} &[R(t_0 + h, t_0 + k) - R(t_0, t_0 + k)] - [R(t_0 + h, t_0) - R(t_0, t_0)] \\ &\quad = k\frac{\partial}{\partial s}[R(t_0 + h, t_0 + \theta_1 k) - R(t_0, t_0 + \theta_1 k)] \end{aligned} \tag{8.5}$$

where $0 < \theta_1 < 1$. Using the existence of $\partial(\partial R/\partial s)/\partial t$ again, we may again apply the Law of the mean to obtain

$$\frac{\partial}{\partial s}[R(t_0 + h, t_0 + \theta_1 k) - R(t_0, t_0 + \theta_1 k)]$$

$$= h \frac{\partial^2}{\partial t \partial s} R(t_0 + \theta_2 h, t_0 + \theta_1 k) \tag{8.6}$$

where $0 < \theta_2 < 1$. Now substitute (8.6) and (8.5) in (8.4) to write

$$\mathscr{E}\left[\frac{\xi(t_0 + h) - \xi(t_0)}{h}\right]\left[\frac{\bar{\xi}(t_0 + k) - \bar{\xi}(t_0)}{k}\right] = \frac{\partial^2}{\partial t \partial s} R(t_0 + \theta_2 h, t_0 + \theta_1 k) \ .$$

By the continuity of $\partial^2 R/\partial t \partial s$ at (t_0, t_0) we conclude that

$$\lim_{h,k \to 0} \mathscr{E}\left[\frac{\xi(t_0 + h) - \xi(t_0)}{h}\right]\left[\frac{\bar{\xi}(t_0 + k) - \bar{\xi}(t_0)}{k}\right] = \frac{\partial^2}{\partial t \partial s} R(t_0, t_0) \ .$$

Hence Loève's criterion (Theorem 6.2 of Chapter I) implies that ξ has a mean square derivative at $t = t_0$ and

$$\mathscr{E}\,|\xi'(t_0)|^2 = \frac{\partial^2}{\partial t \partial s} R(t_0, t_0) = \frac{\partial^2}{\partial s \partial t} R(t_0, t_0) \ . \tag{8.7}$$

The following theorem gives results for differentiability analogous to Theorem 7.2 for mean square continuity.

Theorem 8.7. Let $R(t, s)$ be the covariance function of the mean zero stochastic process $\{\xi(t)\,|\,t \in T\}$. Let $\partial^2 R/\partial t \partial s$ exist on $T \times T$ and be continuous on the diagonal $(t, t) \in T \times T$. Also, let $\partial R/\partial s$ be continuous on $T \times T$. Then $\xi(t)$ has a mean square derivative for all $t \in T$, and $\partial^2 R/\partial t \partial s$ is the covariance function of the derivative process $\{\xi'(t)\,|\,t \in T\}$. Furthermore $\partial^2 R/\partial t \partial s = \partial^2 R/\partial s \partial t$ for all $(t, s) \in T \times T$ and the mixed partial derivatives are continuous on $T \times T$.

Proof. By Theorem 8.6 and the hypotheses on the second mixed partial derivative of R we conclude that $\xi'(t)$ exists in mean square on T. Thus by Theorem 6.3 of Chapter I

$$\lim_{h,k \to 0} \mathscr{E}\left[\frac{\xi(t + h) - \xi(t)}{h}\right]\left[\frac{\bar{\xi}(s + k) - \bar{\xi}(s)}{k}\right] = \mathscr{E}\xi'(t)\bar{\xi}'(s) = K(t, s) \tag{8.8}$$

where $K(t, s)$ is the covariance function of the derivative process $\{\xi'(t)\,|\,t \in T\}$. We must show that

$$K(t, s) = \frac{\partial^2}{\partial t \partial s} R(t, s) \tag{8.9}$$

for all $(t, s) \in T \times T$.

For simplicity in notation let

$$\lambda_{hk} = \mathscr{E}\left[\frac{\xi(t+h)-\xi(t)}{h}\right]\left[\frac{\bar{\xi}(s+k)-\bar{\xi}(s)}{k}\right]$$

$$= \frac{1}{hk}[R(t+h, s+k) - R(t, s+k) - R(t+h, s) + R(t, s)] \quad .$$

The existence of $\partial R/\partial s$ allows us to employ the Law of the mean to write

$$\lambda_{hk} = \frac{1}{h}\frac{\partial}{\partial s}[R(t+h, s+\theta k) - R(t, s+\theta k)]$$

where $0 < \theta < 1$.

From (8.8), as we have noted above,

$$\lim_{h,k\to 0} \lambda_{hk}$$

exists. Also

$$\lim_{k\to 0} \lambda_{hk}$$

exists by the continuity of $\partial R/\partial s$ on $T \times T$. Thus

$$\lim_{h\to 0}\lim_{k\to 0} \lambda_{hk}$$

exists and equals (8.8). But

$$\lim_{h\to 0}\lim_{k\to 0} \lambda_{hk} = \lim_{h\to 0}\frac{\dfrac{\partial}{\partial s}R(t+h, s) - \dfrac{\partial}{\partial s}R(t, s)}{h} = \frac{\partial}{\partial t}\left[\frac{\partial}{\partial s}R(t, s)\right]$$

by definition of the derivative. If we compare this with (8.8) we have the desired result, namely,

$$K(t, s) = \frac{\partial^2}{\partial t \partial s} R(t, s) \quad .$$

Since $\partial^2 R(t, s)/\partial t\partial s$ is a covariance function continuous at every diagonal point $t = s \in T$, we may invoke Theorem 7.2 to conclude that $\partial^2 R/\partial t\partial s$ is continuous on $T \times T$. Thus $\partial^2 R/\partial s\partial t$ also exists and is continuous on $T \times T$, and hence

$$\frac{\partial^2 R}{\partial t\partial s} = \frac{\partial^2 R}{\partial s\partial t}$$

on $T \times T$.

For stationary processes we have the following simplifications.

Corollary 8.2. Let $R(t)$ be the covariance function of the wide-sense stationary stochastic process $\{\xi(t) \mid t \in T\}$. Let $R^{ii}(t)$ exist for all $t \in \text{Dom } R$ and be continuous at $t = 0$. Then $\xi(t)$ has a mean square derivative for all $t \in T$, and $-R^{ii}(t)$ is the covariance function of the derivative process $\{\xi'(t) \mid t \in T\}$. Furthermore, $R^{ii}(t)$ is continuous on Dom R.

Proof. From Theorem 8.7, if $\{\xi(t) \mid t \in T\}$ is weakly stationary, then

$$R(t, s) = R(t - s) \quad .$$

Thus

$$\frac{\partial^2}{\partial t \partial s} R(t, s) = - R^{ii}(t - s)$$

and

$$\frac{\partial^2}{\partial t \partial s} R(t, t) = - R^{ii}(0) \quad ,$$

while the existence of R^{ii} implies the continuity of R'.

By Theorem 5.2, if R is a continuous covariance function and S the corresponding spectral distribution function, then the existence of

$$\int_{-\infty}^{\infty} |f|^2 dS(f)$$

implies the continuity of R^{ii}. Using this fact and Corollary 8.2 we conclude that

Corollary 8.3. Let the spectral distribution function S of the wide-sense stationary process $\{\xi(t) \mid -\infty < t < \infty\}$ exist. Then if

$$\int_{-\infty}^{\infty} |f|^2 dS(f) < \infty \quad ,$$

$\{\xi(t)\}$ is mean square differentiable on $(-\infty, \infty)$.

In fact, summarizing Theorems 6.4, 6.5, 8.4, 8.5 and Corollaries 7.3, 8.3 we have

Corollary 8.4. Let the spectral distribution function S of the real wide-sense stationary process $\{x(t) \mid -\infty < t < \infty\}$ exist. Let

$$\mu(\gamma) = \int_{-\infty}^{\infty} |f|^{\gamma} dS(f) \quad .$$

Then if $\mu(\gamma)$ exists for

(i) $\gamma \geqq 0$, $\{x(t)\}$ is mean square continuous

(ii) $\gamma > 0$, there exists an equivalent process which possesses continuous sample functions with probability one if $\{x(t)\}$ is Gaussian

(iii) $\gamma > 1$, there exists an equivalent process which possesses continuous sample functions with probability one

(iv) $\gamma \geqq 2$, $\{x(t)\}$ has a mean square derivative

(v) $\gamma > 2$, there exists an equivalent process which possesses continuous sample derivatives with probability one if $\{x(t)\}$ is Gaussian

(vi) $\gamma > 3$, there exists an equivalent process which possesses continuous sample derivatives with probability one.

The relationship between mean square continuity and mean square differentiability is the expected one:

Corollary 8.5. If the mean zero stochastic process $\{\xi(t) \mid t \in T\}$ has a mean square derivative on T, then $\{\xi(t)\}$ is continuous in mean square on T.

Proof. By hypothesis

$$\lim_{h \to 0} \mathscr{E} \left| \frac{\xi(t+h) - \xi(t)}{h} \right|^2 \quad , \qquad h \neq 0 \quad ,$$

exists. Therefore

$$\lim_{h \to 0} \mathscr{E} |\xi(t+h) - \xi(t)|^2 = 0 \quad .$$

Note that by Theorems 7.1 and 7.2 the corollary above implies that if $\{\xi(t) \mid t \in T\}$ has a mean square derivative on T, then the covariance function $R(t, s)$ of $\{\xi(t)\}$ is continuous on $T \times T$.

Let $\{\xi(t) \mid t \in T\}$ be a mean zero complex stochastic process with continuous covariance function $R(t, s)$. Suppose $\partial R/\partial t$ exists on $T \times T$. Then from the identity

$$\mathscr{E}\left[\frac{\xi(t+h) - \xi(t)}{h}\right] \bar{\xi}(s) = \frac{R(t+h, s) - R(t, s)}{h}$$

we may write

$$\lim_{h \to 0} \mathscr{E}\left[\frac{\xi(t+h) - \xi(t)}{h}\right] \bar{\xi}(s) = \frac{\partial}{\partial t} R(t, s)$$

by definition of the derivative. However, we may not write

$$\mathscr{E}\xi'(t)\bar{\xi}(s) = \frac{\partial}{\partial t} R(t, s) \tag{8.10}$$

since we have no assurance that the derivative process $\{\xi'(t) \mid t \in T\}$ exists (in any sense). However, if we adopt the hypotheses of Theorem 8.7, then $\{\xi'(t)\}$ *does* exist in mean square and (8.10) is indeed true. Thus we have the following corollary.

Corollary 8.6. Let $R(t, s)$ be the covariance function of the mean zero stochastic process $\{\xi(t) \mid t \in T\}$. Then under the hypotheses of Theorem 8.7, the cross-covariance function of the mean square derivative process $\{\xi'(t)\}$ and the original process $\{\xi(t)\}$ is

$$\frac{\partial}{\partial t} R(t, s) = \mathscr{E}\xi'(t)\bar{\xi}(s) \quad . \tag{8.11}$$

For weakly stationary processes we deduce from Corollaries 8.2 and 8.6 that:

Corollary 8.7. Let $\{\xi(t) \mid t \in T\}$ be a wide-sense stationary stochastic process with covariance function R. Let $R^{ii}(t)$ exist for all $t \in \text{Dom } R$ and be continuous at the origin. Then R' is the cross-covariance function of the mean square derivative process $\{\xi'(t)\}$ and the original process $\{\xi(t)\}$,

$$R'(h) = \mathscr{E}\xi'(t+h)\bar{\xi}(t) \quad .$$

Furthermore, $R'(t)$ is continuous for all $t \in \text{Dom } R$ and

$$R'(t) = -\bar{R}'(-t) \quad .$$

9. Some Miscellaneous Results

Let us examine the properties of sample functions in certain elementary but useful cases. We consider two covariance functions of practical significance in Examples 9.1 and 9.2.

Example 9.1 Let

$$R(t) = e^{-\gamma|t|} \quad , \qquad \gamma > 0 \quad ,$$

be the covariance function of a real wide-sense stationary process $\{x(t) \mid -\infty$

$< t < \infty\}$. Then R is continuous but not differentiable and

$$R(0) - R(h) = \gamma|h| - \tfrac{1}{2}\gamma^2|h|^2 + 0(|h|^3) \quad ,$$

$$6R(0) - 8R(h) + 2R(2h) = 4\gamma|h| + 0(|h|^3)$$

as h approaches zero.

The spectral distribution function is

$$S(f) = \frac{1}{2} + \frac{1}{\pi} \arctan \frac{2\pi f}{\gamma}$$

and the spectral density is

$$W(f) = S'(f) = \frac{2\gamma}{4\pi^2 f^2 + \gamma^2} \quad .$$

The moments of the spectrum are

$$\int_{-\infty}^{\infty} |f|^\alpha dS(f) = \left(\frac{\gamma}{2\pi}\right)^\alpha \sec \tfrac{1}{2}\alpha\pi \quad , \qquad |\alpha| < 1 \quad .$$

Thus by Theorem 6.3 or Theorem 6.5 (or Corollary 8.4) we see that if $\{x(t)\}$ is Gaussian, then there exists an equivalent process which possesses continuous sample functions with probability one. In any event $\{x(t)\}$ is mean square continuous by Corollary 7.1 or Corollary 7.3 (or Corollary 8.4). We also note that for $h \neq 0$

$$\mathscr{E} \frac{x(t+h) - x(t)}{h} = 0$$

while

$$\operatorname{Var} \frac{x(t+h) - x(t)}{h} = \frac{2[R(0) - R(h)]}{h^2} = \frac{2\gamma}{|h|} - \gamma^2 + 0(|h|)$$

—which increases without limit as h approaches zero.

Next we consider a complex density function which is analytic.

Example 9.2 Let

$$R(t) = \exp(-2\pi^2 b^2 t^2) \exp(i2\pi\beta t) \quad , \qquad b > 0 \quad ,$$

be the covariance function of the complex wide-sense stationary process $\{\xi(t) | -\infty < t < \infty\}$. Then R is analytic and

$$R^{(n)}(t) = (-2\pi b)^n R(t) H_n \left(2\pi b t - \frac{i\beta}{b}\right)$$

where $H_n(\cdot)$ is the Hermite polynomial of degree n. Also

$$2R(0) - R(h) - R(-h) = 4\pi^2(b^2 + \beta^2)h^2 + 0(h^4) \quad ,$$

$$6R(0) - 4R(h) - 4R(-h) + R(2h) + R(-2h)$$
$$= 16\pi^4(3b^4 + 6b^2\beta^2 + \beta^4)h^4 + 0(h^6)$$

as h approaches zero.

The spectral distribution function is

$$S(f) = \int_{-\infty}^{f} W(g)dg$$

where

$$W(f) = \frac{1}{b\sqrt{2\pi}} \exp\left[\frac{-(f-\beta)^2}{2b^2}\right]$$

is the spectral density. The moments of W are

$$\int_{-\infty}^{\infty} |f|^\alpha dS(f) = \frac{b^\alpha \Gamma(1+\alpha)}{\sqrt{2\pi}} \exp\left(\frac{-\beta^2}{4b^2}\right)\left[D_{-(1+\alpha)}\left(\frac{\beta}{b}\right) + D_{-(1+\alpha)}\left(-\frac{\beta}{b}\right)\right]$$

for $\alpha > -1$, where $D_{-\nu}(\cdot)$ is the parabolic cylinder function.

Thus by Corollary 8.1 or Theorem 8.5 we see that if $\{\xi(t)\}$ is Gaussian, then there exists an equivalent process which possesses continuous sample derivatives with probability one. In any event $\{\xi(t)\}$ is mean square differentiable by Corollaries 8.2 and 8.3.

While $\exp(-h^2)$ is a covariance function, it is interesting to note that $\exp(-h^4)$ is *not* a covariance function. In fact, we have the following proposition.

Theorem 9.1. For some $\alpha > 0$ let $r^{ii}(h)$ exist for all h where

$$r(h) = 1 - 0(|h|^{2+\alpha}) \quad ,$$

as h approaches zero. Then $r(h)$ cannot be a covariance function.

Proof. The proof is by contradiction. Suppose $r(h)$ were a covariance function. Since $r^{ii}(h)$ is continuous at $h = 0$, Corollary 8.2 implies that

$$\psi(h) = -r^{ii}(h)$$

is a covariance function. Thus

$$|\psi(h)| \leqq \psi(0) \quad ,$$

But $r^{ii}(0) = 0$. Thus $\psi(0) = 0$, and $\psi(h)$ is identically zero.

If we note that

$$\exp(-h^4) = 1 - h^4 + 0(h^8)$$

as h approaches zero, we may apply Theorem 9.1 to conclude that $\exp(-h^4)$ cannot be a covariance function. One may also give a direct proof: Recall that a function $K(t)$ with domain I is said to be nonnegative definite if $K(t) = \bar{K}(-t)$ for all $t \in I$ and

$$\sigma_n(K) = \sum_{j,k=1}^{n} K(t_j - t_k)\, \lambda_j \bar{\lambda}_k \geqq 0$$

for all complex numbers λ_j, $1 \leqq j \leqq n$; all sets of distinct points t_j, $1 \leqq j \leqq n$, in I; and all positive integers n. Now let

$$K(t) = \exp(-t^4) \quad ,$$

$n = 3$, $\lambda_1 = 1$, $\lambda_2 = -1$, $\lambda_3 = 1$, $t_1 = 1$, $t_2 = 1 + \gamma$, $t_3 = 1 + 2\gamma$ (for some $\gamma > 0$). Then

$$\sigma_3(\exp(-t^4)) = 3 - 4\exp(-\gamma^4) + 2\exp(-16\gamma^4)$$

and if, for example, $\gamma^4 = 0.1$, we have $\sigma_3 < 0$. Thus $\exp(-t^4)$ is indefinite and hence cannot be a covariance function.

In the light of Theorem 9.1 we see why we required λ to be positive in Theorems 8.2 and 8.3.

Let $\{\xi(t) | t \in T\}$ be a complex stochastic process. Then $\xi'(t)$ may not exist, or it may exist in mean square, or it may exist with probability one. Suppose now that the real and imaginary parts of $\{\xi(t)\}$ satisfy the conditions of Theorem 8.1. Let $\{h_n\}$ be any sequence of nonzero real numbers converging to zero, and let

$$\lambda_n = \frac{\xi(t + h_n) - \xi(t)}{h_n} \quad .$$

Then [see (4.9) of Chapter I]

$$\lim_{n\to\infty} \lambda_n = \mu$$

and we write

$$\mu = [\xi'(t)]_{\text{ae}} \quad .$$

If $\{\xi(t) | t \in T\}$ has a mean square derivative (Theorem 8.7) on T, then [see (6.4) of Chapter I]

$$\underset{n\to\infty}{\text{l.i.m.}}\ \lambda_n = \nu$$

and we write

$$\nu = [\xi'(t)]_{\mathrm{ms}} \quad .$$

The existence of μ does not imply the existence of ν. The existence of ν does not imply the existence of μ. However, if both μ and ν exist, then they are equal with probability one (Corollary 7.1 of Chapter I). Thus we see that if both μ and ν exist, the subscripts "ae" and "ms" are superfluous. If only one of them exists, the notation $\xi'(t)$ (without subscripts) is unambiguous. We shall give an interesting example later (Theorem 9.3) where the derivative exists a.e. but not in mean square.

As a prelude to Theorem 9.3 we first wish to show that (under suitable hypotheses) the derivative of a Gaussian process is normally distributed. Toward this end let us consider a mean zero complex stationary Gaussian process $\{\xi(t) \mid -\infty < t < \infty\}$ with continuous covariance function R and spectral distribution function S such that

$$\int_{-\infty}^{\infty} |f|^{2+\alpha} dS(f) < \infty$$

for some $\alpha > 0$. Then Theorem 8.5 implies that $[\xi'(t)]_{\mathrm{ae}}$ exists and Corollary 8.3 implies that $[\xi'(t)]_{\mathrm{ms}}$ exists. Thus $x'(t)$ and $y'(t)$ where

$$\xi(t) = x(t) + iy(t)$$

exist a.e. and in mean square. We shall now demonstrate that $\{x'(t)\}$ and $\{y'(t)\}$ are normally distributed.

By Theorem 5.2 we conclude that $R^{ii}(h)$ is continuous for all h, and hence so are $C^{ii}(h)$ and $D^{ii}(h)$ where

$$R = 2C + i2D \quad .$$

As we observed in the discussion preceding Corollary 6.1,

$$C = \tfrac{1}{2} \operatorname{Re} R$$

is the covariance function of the $\{x(t)\}$ and $\{y(t)\}$ processes. By Taylor's theorem

$$C(h) = C(0) + C^{ii}(\tau)\left(\frac{h^2}{2}\right)$$

where τ lies between 0 and h. [Since $C(h) = C(-h)$, we have $C'(0) = 0$.] Also, since C^{ii} is continuous,

$$\lim_{h \to 0} C^{ii}(\tau) = C^{ii}(0) \quad .$$

Now let $\{h_n\}$ be a sequence of nonzero real numbers converging to zero and let

$$v_n = \frac{x(t + h_n) - x(t)}{h_n} \quad .$$

Then

$$\lim_{n \to \infty} v_n = x'(t) = \operatorname*{l.i.m.}_{n \to \infty} v_n \quad .$$

That is, $x'(t)$ exists with probability one and in mean square. The distribution function of v_n is

$$F_n(x) = P[v_n \leqq x] = \frac{1}{\sigma_n \sqrt{2\pi}} \int_{-\infty}^{x} \exp\left(\frac{-u^2}{2\sigma_n^2}\right) du$$

where

$$\sigma_n^2 = \operatorname{Var} v_n = \frac{2[C(0) - C(h_n)]}{h_n^2} = -C^{ii}(\tau_n)$$

and τ_n lies between 0 and h_n.

Now

$$\lim_{n \to \infty} F_n(x) = F(x) = \frac{1}{\sigma \sqrt{2\pi}} \int_{-\infty}^{x} \exp\left(\frac{-u^2}{2\sigma^2}\right) du$$

where

$$\sigma^2 = \lim_{n \to \infty} \sigma_n^2 = -C^{ii}(0) \quad .$$

Thus $\{v_n\}$ converges in distribution. Since the convergence of $\{v_n\}$ with probability one (or in mean square) implies convergence in probability, Theorem 5.2 of Chapter I implies that the distribution function of $x'(t)$ is normal $(0, -C^{ii}(0))$. Clearly $y'(t)$ also has this property. Summarizing:

Theorem 9.2. Let $\{\xi(t) \mid -\infty < t < \infty\}$ be a mean zero complex stationary Gaussian process with continuous covariance function R and spectral distribution function S. Let

$$\int_{-\infty}^{\infty} |f|^{2+\alpha} dS(f) < \infty$$

for some $\alpha > 0$. Then the derivative process $\{\xi'(t) \mid -\infty < t < \infty\}$ exists in mean square and a.e., and the real and imaginary components of ξ' are normal $(0, -C^{ii}(0))$ where

$$C = \tfrac{1}{2} \operatorname{Re} R \quad .$$

Using vector processes one can show, for example, that under the hypotheses of Theorem 9.2,

$$\boldsymbol{z}' = \{x(t), y'(t), y(t), x'(t)\}$$

is normal $(\boldsymbol{0}, \boldsymbol{M})$ where

$$\boldsymbol{M} = \begin{bmatrix} C(0) & D'(0) & 0 & 0 \\ D'(0) & -C^{ii}(0) & 0 & 0 \\ 0 & 0 & C(0) & -D'(0) \\ 0 & 0 & -D'(0) & -C^{ii}(0) \end{bmatrix} .$$

After these preliminaries we may now state and prove Theorem 9.3.

Theorem 9.3. Let $\{\xi(t) | -\infty < t < \infty\}$ be a mean zero complex stationary Gaussian process with positive definite covariance function R and corresponding spectral distribution function S. Let $\int_{-\infty}^{\infty} f\, dS(f) \neq 0$, and let

$$\int_{-\infty}^{\infty} |f|^{2+\alpha}\, dS(f) < \infty \tag{9.1}$$

for some $\alpha > 0$. Let $\xi(t) = |\xi(t)| e^{i\theta(t)}$. Then $[\theta'(t)]_{\text{ae}}$ exists, $[\theta'(t)]_{\text{ms}}$ does not exist. Furthermore

$$\mathscr{E}[\theta'(t)]_{\text{ae}} = \frac{2\pi \int_{-\infty}^{\infty} f\, dS(f)}{\int_{-\infty}^{\infty} dS(f)} \quad . \tag{9.2}$$

Proof. Write $\xi(t)$ in polar form as

$$\xi(t) = r(t) e^{i\theta(t)}$$

where $r(t) = |\xi(t)|$. Then

$$\theta'(t) = \frac{x(t)y'(t) - y(t)x'(t)}{x^2(t) + y^2(t)} \quad .$$

Since $x'(t)$ and $y'(t)$ are continuous with probability one (by Theorem 9.2), so is $\theta'(t)$. However, we cannot conclude that $\theta(t)$ is mean square differentiable. In fact we shall show that $[\theta'(t)]_{\text{ms}}$ does not exist.

Toward this end let us compute the probability density function of $\theta'(t)$. Make the change of variable from $\boldsymbol{z}' = \{x, y', y, x'\}$ to r, θ, r', θ' with Jacobian

r^2. Then the joint density function h of r, θ, r', θ' is related to the joint [normal $(\boldsymbol{0}, \boldsymbol{M})$] density function f of $\boldsymbol{z}$ by the equation

$$h(r, \theta, r', \theta') = r^2 f(\boldsymbol{z})$$

$$= \frac{r^2}{(2\pi)^2(\det \boldsymbol{M})^{1/2}} \exp\{-\tfrac{1}{2}[(a_2\theta'^2 + 2b\theta' + a_0)r^2 + a_2 r'^2]\}$$

where

$$\boldsymbol{M}^{-1} = \begin{vmatrix} a_0 & b & 0 & 0 \\ b & a_2 & 0 & 0 \\ 0 & 0 & a_0 & -b \\ 0 & 0 & -b & a_2 \end{vmatrix} .$$

The density function p of θ' is the marginal frequency function

$$p(\theta') = \int_0^{2\pi} d\theta \int_0^{\infty} dr \int_{-\infty}^{\infty} dr'\, h(r, \theta, r', \theta')$$
$$= \frac{C(\det \boldsymbol{M})^{1/2}}{2[(C\theta' - D')^2 + (\det \boldsymbol{M})^{1/2}]^{3/2}} \tag{9.3}$$

where for simplicity in notation we have written $C = C(0)$, $D' = D'(0)$. [The probability density function of r is Rayleigh, θ is uniform, and r' is Gaussian.]

From (9.3) we see that

$$\mathscr{E}[\theta'(t)]_{\text{ae}} = \frac{D'(0)}{C(0)} = \frac{2\pi \int_{-\infty}^{\infty} f\, dS(f)}{\int_{-\infty}^{\infty} dS(f)} \tag{9.4}$$

since

$$R'(0) = 2C'(0) + i2D'(0) = i2D'(0) = i2\pi \int_{-\infty}^{\infty} f\, dS(f)$$

and

$$R(0) = 2C(0) + i2D(0) = 2C(0) = \int_{-\infty}^{\infty} dS(f) .$$

[From Theorem 5.2 we see that (9.1) implies the existence and continuity of $R'(t)$.] We also see that

$$\text{Var}\,[\theta'(t)]_{\text{ae}} = \infty .$$

Since $\mathscr{E}\theta(t) = \pi$, we have

$$\mathscr{E}\frac{\theta(t+h)-\theta(t)}{h} = 0 \quad , \qquad h \neq 0 \quad . \tag{9.5}$$

Now suppose

$$\left\{\frac{\theta(t+h)-\theta(t)}{h}\right\}$$

converged in mean square (to $[\theta'(t)]_{\mathrm{ms}}$) as h approached zero. Then by Loève's criterion (see Corollary 6.1 of Chapter I) we would have

$$\lim_{h\to 0} \mathscr{E}\frac{\theta(t+h)-\theta(t)}{h} = \mathscr{E}[\theta'(t)]_{\mathrm{ms}} \quad . \tag{9.6}$$

But if $\{[\theta(t+h)-\theta(t)]/h\}$ converged in mean square to $[\theta'(t)]_{\mathrm{ms}}$, we would have

$$[\theta'(t)]_{\mathrm{ms}} = [\theta'(t)]_{\mathrm{ae}}$$

with probability one, and hence

$$\mathscr{E}[\theta'(t)]_{\mathrm{ms}} = \mathscr{E}[\theta'(t)]_{\mathrm{ae}} \quad .$$

But by (9.5) and (9.6)

$$\mathscr{E}[\theta'(t)]_{\mathrm{ms}} = 0$$

while by (9.4)

$$\mathscr{E}[\theta'(t)]_{\mathrm{ae}} \neq 0 \quad .$$

III

THE COMPLEX DENSITY FUNCTION

0. Introduction

When discussing real Gaussian processes, we soon encounter the multivariate normal density function. However, when discussing *complex* Gaussian processes, we never mentioned a complex density function. (In fact, the term is almost self-contradictory, since, for example, a density function must be nonnegative.) What we did was to relate the finite dimensional distributions of the complex variates to the joint distributions of their real and imaginary parts. Nevertheless, it is frequently convenient, both theoretically and practically, to have a density function available. For example, if such a function existed, one could calculate a moment generating function and hence the integral moments of a complex Gaussian process. Naturally one could always decompose the random vector into real and imaginary components, and then use real multivariate analysis to determine the moments. However, from our experience in other branches of analysis, this decomposition is frequently laborious and hard to generalize.

In Section 1 we shall define, in a natural fashion, a nonnegative scalar function that may, without stretching the imagination, be defined as a density function of an n-dimensional complex Gaussian vector. Various equivalent forms and marginal distributions will also be derived.

If $\boldsymbol{x}$ is a random vector, and $\boldsymbol{t}$ a vector of the same dimension and independent of $\boldsymbol{x}$, we may compute $\mathscr{E}e^{\boldsymbol{t}'\boldsymbol{x}}$ (if it exists). If $\boldsymbol{t}$ is real, we call $\mathscr{E}e^{\boldsymbol{t}'\boldsymbol{x}}$ the moment generating function of $\boldsymbol{x}$, and if $\boldsymbol{t}$ is pure imaginary, we call $\mathscr{E}e^{\boldsymbol{t}'\boldsymbol{x}}$ the characteristic function of $\boldsymbol{x}$. We shall assume in Section 2 that $\boldsymbol{t}$ is a complex vector and compute $\mathscr{E}e^{\boldsymbol{t}'\boldsymbol{x}}$. In particular, we shall use this result to calculate the integral moments of the various components of a complex Gaussian vector.

Suppose $\boldsymbol{z}' = \{z_1, \cdots, z_n\} = \boldsymbol{x} + i\boldsymbol{y}$ is a complex Gaussian vector. Then the Cartesian components $\boldsymbol{x}$ and $\boldsymbol{y}$ have n-dimensional marginal Gaussian density functions. We may also write $z_k = r_k \exp(i\theta_k)$, $1 \leqq k \leqq n$, and ask what is the distribution of the amplitudes $\boldsymbol{r}' = \{r_1, \cdots, r_n\}$ and phases $\boldsymbol{\theta}' = \{\theta_1, \cdots, \theta_n\}$. This problem is discussed in Section 3, and the results are used in Section 4 to calculate certain nonintegral moments of the amplitudes and some moments of the phases.

1. Complex Density Functions

Let x and y be real random variables, defined on the same probability space, with a joint probability density function $f(x, y)$. Let $z = x + iy$. Then $f(x, y)$ may be written as a function of z, say, $f(x, y) = f(z)$. That is, $f(z)$ is a real-valued function of the complex variable z.

Suppose now that $\boldsymbol{z}$ is an n-dimensional complex random vector with mean vector $\boldsymbol{c}$ and positive definite covariance matrix $\boldsymbol{R}$. That is,

$$\mathscr{E}\boldsymbol{z} = \boldsymbol{c}$$

and

$$\mathscr{E}(\boldsymbol{z} - \boldsymbol{c})(\bar{\boldsymbol{z}} - \bar{\boldsymbol{c}})' = \boldsymbol{R} \quad .$$

Now the quadratic form $(\bar{\boldsymbol{z}} - \bar{\boldsymbol{c}})'\boldsymbol{R}^{-1}(\boldsymbol{z} - \boldsymbol{c})$ is real and $\det \boldsymbol{R} > 0$. By analogy with real multivariate Gaussian theory (see Section 3 of Chapter II) we are invited to consider

$$f(\boldsymbol{z}) = \frac{1}{\pi^n \det \boldsymbol{R}} \exp\left[-(\bar{\boldsymbol{z}} - \bar{\boldsymbol{c}})'\boldsymbol{R}^{-1}(\boldsymbol{z} - \boldsymbol{c})\right] \quad , \tag{1.1}$$

which is a real-valued scalar function of the complex vector $\boldsymbol{z}$.

From Section 2 of Chapter II we may write

$$\begin{aligned}(\bar{\boldsymbol{z}} - \bar{\boldsymbol{c}})'\boldsymbol{R}^{-1}(\boldsymbol{z} - \boldsymbol{c}) &= \tfrac{1}{2}(\boldsymbol{\eta} - \boldsymbol{h})'\boldsymbol{\Sigma}^{-1}(\boldsymbol{\eta} - \boldsymbol{h}) \\ &= \tfrac{1}{2}(\bar{\boldsymbol{\eta}} - \bar{\boldsymbol{h}})'\boldsymbol{\Omega}^{-1}(\boldsymbol{\eta} - \boldsymbol{h}) \\ &= \tfrac{1}{2}(\boldsymbol{\xi} - \boldsymbol{\gamma})'\boldsymbol{M}^{-1}(\boldsymbol{\xi} - \boldsymbol{\gamma})\end{aligned} \tag{1.2}$$

where

$$\boldsymbol{\eta} = \begin{bmatrix} \boldsymbol{z} \\ \bar{\boldsymbol{z}} \end{bmatrix} \quad , \qquad \boldsymbol{h} = \mathscr{E}\boldsymbol{\eta} = \begin{bmatrix} \boldsymbol{c} \\ \bar{\boldsymbol{c}} \end{bmatrix} \tag{1.3a}$$

and

$$\boldsymbol{\xi} = \begin{bmatrix} \boldsymbol{x} \\ \boldsymbol{y} \end{bmatrix} \quad , \qquad \boldsymbol{\gamma} = \mathscr{E}\boldsymbol{\xi} = \begin{bmatrix} \boldsymbol{a} \\ \boldsymbol{b} \end{bmatrix} \tag{1.3b}$$

while

$$\boldsymbol{x} = \operatorname{Re} \boldsymbol{z} \quad , \qquad \boldsymbol{a} = \mathscr{E}\boldsymbol{x} = \operatorname{Re} \boldsymbol{c} \quad ,$$
$$\boldsymbol{y} = \operatorname{Im} \boldsymbol{z} \quad , \qquad \boldsymbol{b} = \mathscr{E}\boldsymbol{y} = \operatorname{Im} \boldsymbol{c} \quad .$$

Then

$$\boldsymbol{\Sigma} = \begin{bmatrix} \boldsymbol{0} & \boldsymbol{R} \\ \bar{\boldsymbol{R}} & \boldsymbol{0} \end{bmatrix} ,$$

$$\boldsymbol{\Omega} = \begin{bmatrix} \boldsymbol{R} & \boldsymbol{0} \\ \boldsymbol{0} & \bar{\boldsymbol{R}} \end{bmatrix} ,$$

and

$$\boldsymbol{M} = \tfrac{1}{2} \operatorname{Re} \begin{bmatrix} \boldsymbol{R} & i\boldsymbol{R} \\ i\bar{\boldsymbol{R}} & \boldsymbol{R} \end{bmatrix} \tag{1.4}$$

where $\boldsymbol{0}$ is the $n \times n$ zero matrix. Thus

$$\begin{aligned} \det \boldsymbol{\Sigma} &= (-1)^n (\det \boldsymbol{R})^2 \quad , \\ \det \boldsymbol{\Omega} &= (\det \boldsymbol{R})^2 \quad , \\ \det \boldsymbol{M} &= 2^{-2n} (\det \boldsymbol{R})^2 \quad . \end{aligned} \tag{1.5}$$

By reference to the formulas above we see that (1.1) may also be written as

$$f(\boldsymbol{z}) = \frac{1}{\pi^n (\det \boldsymbol{\Omega})^{1/2}} \exp\left[-\tfrac{1}{2}(\bar{\boldsymbol{\eta}} - \bar{\boldsymbol{h}})' \boldsymbol{\Omega}^{-1} (\boldsymbol{\eta} - \boldsymbol{h})\right] \quad . \tag{1.6}$$

or as

$$f(\boldsymbol{z}) = \frac{1}{(2\pi)^n (\det \boldsymbol{M})^{1/2}} \exp\left[-\tfrac{1}{2}(\boldsymbol{\xi} - \boldsymbol{\gamma})' \boldsymbol{M}^{-1} (\boldsymbol{\xi} - \boldsymbol{\gamma})\right] \quad . \tag{1.7}$$

Hence from (1.7) [cf. (3.5) of Chapter II], we see that $f(\boldsymbol{z})$ is nonnegative and

$$\int_{-\infty}^{\infty} \int_{-\infty}^{\infty} f(\boldsymbol{z})\, d\boldsymbol{x}\, d\boldsymbol{y} = 1 \quad . \tag{1.8}$$

Since $f(\boldsymbol{z}) \geqq 0$, and its integral of over all of real $2n$-dimensional Euclidean space is unity, it is a density function. If an n-dimensional complex random vector $\boldsymbol{z}$ has the density function (1.1), then we shall say that $\boldsymbol{z}$ has a *joint n-dimensional complex Gaussian density function* (see [Go], [Mil 2], [Wo]) or that $\boldsymbol{z}$ is *normally distributed* ($\boldsymbol{c}$, $\boldsymbol{R}$) in n complex dimensions.

We see from (1.7) that $\mathscr{E}\boldsymbol{\xi} = \boldsymbol{\gamma}$, from which follow $\mathscr{E}\boldsymbol{\eta} = \boldsymbol{h}$ and $\mathscr{E}\boldsymbol{z} = \boldsymbol{c}$. Second moments are also available by inspection. In particular, from (1.7) and (1.4)

$$\mathscr{E}(\boldsymbol{x}-\boldsymbol{a})(\boldsymbol{x}-\boldsymbol{a})' = \tfrac{1}{2}\,\mathrm{Re}\,\boldsymbol{R} = \mathscr{E}(\boldsymbol{y}-\boldsymbol{b})(\boldsymbol{y}-\boldsymbol{b})' \tag{1.9a}$$

and

$$\mathscr{E}(\boldsymbol{x}-\boldsymbol{a})(\boldsymbol{y}-\boldsymbol{b})' = -\tfrac{1}{2}\mathrm{Im}\,\boldsymbol{R} = -\,\mathscr{E}(\boldsymbol{y}-\boldsymbol{b})(\boldsymbol{x}-\boldsymbol{a})' \quad . \tag{1.9b}$$

Hence we conclude that

$$\mathscr{E}(\boldsymbol{z}-\boldsymbol{c})(\bar{\boldsymbol{z}}-\bar{\boldsymbol{c}})' = \boldsymbol{R}$$

and

$$\mathscr{E}(\boldsymbol{z}-\boldsymbol{c})(\boldsymbol{z}-\boldsymbol{c})' = \boldsymbol{0} \quad .$$

Compare this last formula with the condition (4.3) of Chapter II.

Let us now calculate the marginal density functions. Let p and q be non-negative integers whose sum is n, and partition $\boldsymbol{x}$, $\boldsymbol{y}$, and $\boldsymbol{M}$ as

$$\boldsymbol{x} = \begin{bmatrix} \boldsymbol{x}_p \\ \boldsymbol{x}_q \end{bmatrix} \quad , \qquad \boldsymbol{y} = \begin{bmatrix} \boldsymbol{y}_p \\ \boldsymbol{y}_q \end{bmatrix} \quad ,$$

where $\boldsymbol{x}_p$ and $\boldsymbol{y}_p$ are p-dimensional vectors and $\boldsymbol{x}_q$ and $\boldsymbol{y}_q$ are q-dimensional vectors, while

$$\boldsymbol{M} = \begin{bmatrix} \boldsymbol{M}_{pp} & \boldsymbol{M}_{pq} \\ \boldsymbol{M}_{qp} & \boldsymbol{M}_{qq} \end{bmatrix} \quad ,$$

where $\boldsymbol{M}_{pp}$ is a $2p \times 2p$ matrix, $\boldsymbol{M}_{pq}$ is a $2p \times 2q$ matrix, $\boldsymbol{M}_{qp}$ is a $2q \times 2p$ matrix, and $\boldsymbol{M}_{qq}$ is a $2q \times 2q$ matrix. Then [see (3.6) of Chapter II] from (1.7)

$$\begin{aligned} f_p(\boldsymbol{z}_p) &= \int_{-\infty}^{\infty} \int_{-\infty}^{\infty} f(\boldsymbol{z})\, d\boldsymbol{x}_q\, d\boldsymbol{y}_q \\ &= \frac{1}{(2\pi)^p\,(\det \boldsymbol{M}_{pp})^{1/2}} \exp\left[-\tfrac{1}{2}(\boldsymbol{\xi}_p - \boldsymbol{\gamma}_p)'\boldsymbol{M}_{pp}^{-1}(\boldsymbol{\xi}_p - \boldsymbol{\gamma}_p)\right] \end{aligned}$$

where $\boldsymbol{\xi}_p$ is the partitioned vector

$$\boldsymbol{\xi}_p = \begin{bmatrix} \boldsymbol{x}_p \\ \boldsymbol{y}_p \end{bmatrix}$$

and

$$\begin{aligned} \boldsymbol{\gamma}_p &= \mathscr{E}\boldsymbol{\xi}_p \quad , \\ \boldsymbol{M}_{pp} &= \mathscr{E}(\boldsymbol{\xi}_p - \boldsymbol{\gamma}_p)(\boldsymbol{\xi}_p - \boldsymbol{\gamma}_p)' \quad . \end{aligned}$$

(Recall that $\boldsymbol{M}_{pp}$ is a $2p \times 2p$ matrix.) It therefore follows that

$$f_p(\boldsymbol{z}_p) = \frac{1}{\pi^p \det \boldsymbol{R}_{pp}} \exp\left[-(\bar{\boldsymbol{z}}_p - \bar{\boldsymbol{c}}_p)'\boldsymbol{R}_{pp}^{-1}(\boldsymbol{z}_p - \boldsymbol{c}_p)\right]$$

is a p-dimensional complex Gaussian frequency function where $\boldsymbol{z}_p = \boldsymbol{x}_p + i\boldsymbol{y}_p$ and $\boldsymbol{c}_p = \mathscr{E}\boldsymbol{z}_p = \boldsymbol{a}_p + i\boldsymbol{b}_p$. The matrix $\boldsymbol{R}_{pp}$ is the $p \times p$ submatrix in the upper left-hand corner of $\boldsymbol{R}$, and

$$\boldsymbol{R}_{pp} = \mathscr{E}(\boldsymbol{z}_p - \boldsymbol{c}_p)(\bar{\boldsymbol{z}}_p - \bar{\boldsymbol{c}}_p)' \quad .$$

One may also write

$$f_p(\boldsymbol{z}_p) = \frac{1}{\pi^p (\det \boldsymbol{\Omega}_{pp})^{1/2}} \exp\left[-\tfrac{1}{2}(\bar{\boldsymbol{\eta}}_p - \bar{\boldsymbol{h}}_p)'\boldsymbol{\Omega}_{pp}^{-1}(\boldsymbol{\eta}_p - \boldsymbol{h}_p)\right]$$

where $\boldsymbol{\eta}_p$, $\boldsymbol{h}_p$, and $\boldsymbol{\Omega}_{pp}$ are defined in the expected fashion.

With suitable modifications, almost every result in real multivariate Gaussian theory has a complex analog. For example, if we write

$$f(\boldsymbol{z}\,|\,\boldsymbol{c}, \boldsymbol{R}) = \frac{1}{\pi^n \det \boldsymbol{R}} \exp\left[-(\bar{\boldsymbol{z}} - \bar{\boldsymbol{c}})'\boldsymbol{R}^{-1}(\boldsymbol{z} - \boldsymbol{c})\right] \quad ,$$

then

$$\int_{-\infty}^{\infty}\int_{-\infty}^{\infty} f(\boldsymbol{z}\,|\,\boldsymbol{c}, \boldsymbol{R}) f(\boldsymbol{z}\,|\,\boldsymbol{d}, \boldsymbol{S})\, d\boldsymbol{x}\, d\boldsymbol{y} = f(\boldsymbol{c}\,|\,\boldsymbol{d}, \boldsymbol{R} + \boldsymbol{S})$$

where $\boldsymbol{R}$ and $\boldsymbol{S}$ are positive definite. And the Bhattacharyya coefficient

$$\rho = \int_{-\infty}^{\infty}\int_{-\infty}^{\infty} [f(\boldsymbol{z}\,|\,\boldsymbol{c}, \boldsymbol{R}) f(\boldsymbol{z}\,|\,\boldsymbol{d}, \boldsymbol{S})]^{1/2}\, d\boldsymbol{x}\, d\boldsymbol{y}$$

is

$$\rho = \frac{2^n[(\det \boldsymbol{R})(\det \boldsymbol{S})]^{1/2}}{\det(\boldsymbol{R} + \boldsymbol{S})} \exp\left[-\tfrac{1}{2}(\bar{\boldsymbol{c}} - \bar{\boldsymbol{d}})'(\boldsymbol{R} + \boldsymbol{S})^{-1}(\boldsymbol{c} - \boldsymbol{d})\right] \quad .$$

We also have the important result that if $\boldsymbol{z}' = \{z_1, \cdots, z_n\}$ is normal $(\boldsymbol{c}, \boldsymbol{R})$, and if $\boldsymbol{\gamma}' = \{\gamma_1, \cdots, \gamma_n\}$ is a complex constant vector, then

$$\zeta = \gamma_1 z_1 + \cdots + \gamma_n z_n = \boldsymbol{\gamma}'\boldsymbol{z} \tag{1.10}$$

is normally distributed with mean

$$\mathscr{E}\zeta = \mathscr{E}\boldsymbol{\gamma}'\boldsymbol{z} = \boldsymbol{\gamma}'\boldsymbol{c} \tag{1.11}$$

and variance

$$\operatorname{Var}\zeta = \mathscr{E}|\zeta - \mathscr{E}\zeta|^2$$

$$= \mathscr{E}(\boldsymbol{\gamma}'\boldsymbol{z} - \boldsymbol{\gamma}'\boldsymbol{c})(\bar{\boldsymbol{z}}'\bar{\boldsymbol{\gamma}} - \bar{\boldsymbol{c}}'\bar{\boldsymbol{\gamma}})$$

$$= \boldsymbol{\gamma}'\boldsymbol{R}\bar{\boldsymbol{\gamma}} \quad . \tag{1.12}$$

Finally, we shall establish a formula [(1.13)] which is useful in the proof of *Price's theorem* (see [McG]). Let $\boldsymbol{z}$ be normally distributed $(\boldsymbol{c}, \boldsymbol{R})$ in n complex dimensions where $\boldsymbol{R} = \| R_{jk} \|$ is positive definite. We shall consider $z_1, \cdots, z_n, \bar{z}_1, \cdots, \bar{z}_n$ as $2n$ independent variables, and likewise assume $R_{jj}, R_{jk}, j \neq k, 1 \leqq j, k \leqq n$, to be n^2 independent variables. Then if f is the density function of $\boldsymbol{z}$, we shall prove that

$$\frac{\partial^2 f(\boldsymbol{z})}{\partial z_j \partial \bar{z}_k} = \frac{\partial f(\boldsymbol{z})}{\partial R_{jk}} \quad . \tag{1.13}$$

Now

$$f(\boldsymbol{z}) = \frac{1}{\pi^n \det \boldsymbol{R}} e^{-q(\boldsymbol{z})}$$

where

$$q(\boldsymbol{z}) = (\bar{\boldsymbol{z}} - \bar{\boldsymbol{c}})'\boldsymbol{R}^{-1}(\boldsymbol{z} - \boldsymbol{c}) = \sum_{\alpha, \beta=1}^{n} S_{\alpha\beta}(\bar{z}_\alpha - \bar{c}_\alpha)(z_\beta - c_\beta)$$

and

$$\boldsymbol{R}^{-1} = \| S_{\alpha\beta} \| \quad .$$

It easily follows that

$$\frac{\partial^2 f(\boldsymbol{z})}{\partial z_j \partial \bar{z}_k} = -\left[S_{kj} - \frac{\partial q(\boldsymbol{z})}{\partial z_j} \frac{\partial q(\boldsymbol{z})}{\partial \bar{z}_k} \right] f(\boldsymbol{z}) \quad , \tag{1.14}$$

and since

$$\frac{\partial}{\partial R_{jk}} (\det \boldsymbol{R}) = (\det \boldsymbol{R})\, S_{kj} \quad ,$$

we have

$$\frac{\partial f(\boldsymbol{z})}{\partial R_{jk}} = -\left[S_{kj} + \frac{\partial q(\boldsymbol{z})}{\partial R_{jk}} \right] f(\boldsymbol{z}) \quad . \tag{1.15}$$

Comparing (1.14) and (1.15), we see that to prove (1.13) it is sufficient to show that

$$\frac{\partial q(\boldsymbol{z})}{\partial R_{jk}} = -\frac{\partial q(\boldsymbol{z})}{\partial z_j} \frac{\partial q(\boldsymbol{z})}{\partial \bar{z}_k} \quad . \tag{1.16}$$

Trivially,

$$\frac{\partial q(\mathbf{z})}{\partial z_j}\frac{\partial q(\mathbf{z})}{\partial \bar{z}_k} = \sum_{\alpha,\beta=1}^{n} S_{\alpha j}\, S_{k\beta}(\bar{z}_\alpha - \bar{c}_\alpha)(z_\beta - c_\beta) \tag{1.17}$$

and

$$\frac{\partial q(\mathbf{z})}{\partial R_{jk}} = \sum_{\alpha,\beta=1}^{n} \frac{\partial S_{\alpha\beta}}{\partial R_{jk}}(\bar{z}_\alpha - \bar{c}_\alpha)(z_\beta - c_\beta) \quad . \tag{1.18}$$

Now it is not difficult to show that

$$\frac{\partial S_{\alpha\beta}}{\partial R_{jk}} = -S_{k\beta}S_{\alpha j} \quad . \tag{1.19}$$

(In fact, (1.19) is essentially a special case of *Jacobi's theorem,* see [Ai, page 98].) Substituting (1.19) in (1.18) and comparing with (1.17) we see the truth of (1.16), and *a fortiori,* of (1.13).

2. Generating Functions

Let $\{z(t)\,|\,t \in T\}$ be a complex Gaussian process with mean zero and covariance function R. Then for all t, s in T,

$$\mathscr{E}z(t)\,\bar{z}(s) = R(t, s)$$

$$\mathscr{E}z(t)\,z(s) = 0$$

$$\mathscr{E}x(t)\,x(s) = \tfrac{1}{2}\,\mathrm{Re}\;R(t, s) = \mathscr{E}y(t)\,y(s)$$

$$\mathscr{E}x(t)\,y(s) = -\tfrac{1}{2}\mathrm{Im}\;R(t, s) = -\mathscr{E}x(s)\,y(t)$$

where $z = x + iy$ [see (4.2), (4.3), (6.5), (6.7), (6.8) of Chapter II and (1.9)]. Now using the properties of real Gaussian processes, we may compute various moments. For example,

$$\mathscr{E}z(t)\,\bar{z}(s)\,z(t')\,\bar{z}(s') = R(t, s)\,R(t', s') + R(t, s')\,R(t', s) \tag{2.1}$$

and

$$\mathscr{E}z(t)\,z(s)\,z(t')\,z(s') = 0 = \mathscr{E}z(t)\,z(s)\,z(t')\,\bar{z}(s') \tag{2.2}$$

for all t, s, t', s' in T. The computations above are laborious. We shall systematize the calculation of moments by computing a moment generating function.

We begin by computing $\mathscr{E}e^{\mathbf{t}'\boldsymbol{\xi}}$ where $\boldsymbol{\xi}$ has been defined by (1.3b) and $\mathbf{t}$ is a $2n$-dimensional vector independent of $\boldsymbol{\xi}$. If $\mathbf{t}$ is real, then $\mathscr{E}e^{\mathbf{t}'\boldsymbol{\xi}}$ is the *moment generating function* of $\boldsymbol{\xi}$. If $\mathbf{t}$ is pure imaginary, then $\mathscr{E}e^{\mathbf{t}'\boldsymbol{\xi}}$ is the *characteristic function* of $\boldsymbol{\xi}$. We shall find $\mathscr{E}e^{\mathbf{t}'\boldsymbol{\xi}}$ when $\mathbf{t}$ is an arbitrary complex vector. Thus both moment generating function and characteristic function will be subsumed under our result.

Theorem 2.1. Let $\boldsymbol{\xi}$ be normally distributed $(\boldsymbol{\gamma}, \boldsymbol{M})$ in $2n$ real dimensions and let $\boldsymbol{M}$ be positive definite. Let $\boldsymbol{t}' = \{t_1, \cdots, t_{2n}\}$ be an arbitrary $2n$-dimensional complex vector independent of $\boldsymbol{\xi}$. Then

$$\mathscr{E} \exp(\boldsymbol{t}'\boldsymbol{\xi}) = \exp(\tfrac{1}{2}\boldsymbol{t}'\boldsymbol{M}\boldsymbol{t} + \boldsymbol{t}'\boldsymbol{\gamma}) \quad . \tag{2.3}$$

Proof. By definition [see (1.7)]

$$\mathscr{E} \exp(\boldsymbol{t}'\boldsymbol{\xi}) = \frac{1}{(2\pi)^n (\det \boldsymbol{M})^{1/2}} \int_{-\infty}^{\infty} \exp(\boldsymbol{t}'\boldsymbol{\xi}) \exp[-\tfrac{1}{2}(\boldsymbol{\xi} - \boldsymbol{\gamma})'\boldsymbol{M}^{-1}(\boldsymbol{\xi} - \boldsymbol{\gamma})]\, d\boldsymbol{\xi}.$$

Since $\boldsymbol{M}$ is positive definite, there exists a real nonsingular matrix $\boldsymbol{Q}$ such that $\boldsymbol{M} = \boldsymbol{Q}\boldsymbol{Q}'$. Now make the change of variable $\boldsymbol{\xi} = \boldsymbol{Q}\boldsymbol{\zeta} + \boldsymbol{\gamma}$ with Jacobian $|\det \boldsymbol{Q}|$. Then

$$\mathscr{E} \exp(\boldsymbol{t}'\boldsymbol{\xi}) = (2\pi)^{-n} \exp(\boldsymbol{t}'\boldsymbol{\gamma}) \int_{-\infty}^{\infty} \exp(\boldsymbol{\zeta}'\boldsymbol{q}) \exp[-\tfrac{1}{2}\boldsymbol{\zeta}'\boldsymbol{\zeta}]\, d\boldsymbol{\zeta} \tag{2.4}$$

where $\boldsymbol{q} = \boldsymbol{Q}'\boldsymbol{t}$.

Now if $c = a + ib$ we may write the one-dimensional integral

$$\int_{-\infty}^{\infty} \exp(cu) \exp(-\tfrac{1}{2}u^2)\, du$$

as

$$\exp(\tfrac{1}{2}a^2) \exp(iab) \int_{-\infty}^{\infty} \exp(ibv) \exp(-\tfrac{1}{2}v^2)\, dv = \sqrt{2\pi} \exp(\tfrac{1}{2}c^2)$$

and from (2.4)

$$\mathscr{E} \exp(\boldsymbol{t}'\boldsymbol{\xi}) = \exp(\boldsymbol{t}'\boldsymbol{\gamma}) \exp(\tfrac{1}{2}\boldsymbol{q}'\boldsymbol{q}) \quad .$$

But $\boldsymbol{q}'\boldsymbol{q} = \boldsymbol{t}'\boldsymbol{Q}\boldsymbol{Q}'\boldsymbol{t} = \boldsymbol{t}'\boldsymbol{M}\boldsymbol{t}$, and our theorem is proved.

If we let

$$\boldsymbol{t}' = i\boldsymbol{s}' = i\{s_1, \cdots, s_{2n}\}$$

where $\boldsymbol{s}$ is a real vector, then from (2.3) the characteristic function ϕ of $\boldsymbol{\xi}$ is

$$\phi(\boldsymbol{s}) = \mathscr{E} \exp(i\boldsymbol{s}'\boldsymbol{\xi}) = \exp(-\tfrac{1}{2}\boldsymbol{s}'\boldsymbol{M}\boldsymbol{s} + i\boldsymbol{s}'\boldsymbol{\gamma}) \quad .$$

In terms of complex quantities,

$$\phi(\boldsymbol{\sigma}) = \mathscr{E} \exp(i \operatorname{Re} \bar{\boldsymbol{\sigma}}'\boldsymbol{z}) = \exp(-\tfrac{1}{4}\, \bar{\boldsymbol{\sigma}}'\boldsymbol{R}\boldsymbol{\sigma} + i \operatorname{Re} \bar{\boldsymbol{\sigma}}'\boldsymbol{c})$$

where $\boldsymbol{z}$, $\boldsymbol{c}$, $\boldsymbol{R}$ are related to $\boldsymbol{\xi}$, $\boldsymbol{\gamma}$, $\boldsymbol{M}$ by (1.3b), (1.4); and

$$\boldsymbol{\sigma} = \begin{bmatrix} s_1 + is_{n+1} \\ s_2 + is_{n+2} \\ \cdot \quad \cdot \quad \cdot \\ s_n + is_{2n} \end{bmatrix} \quad .$$

If we let

$$\boldsymbol{t} = \begin{bmatrix} \boldsymbol{u} \\ i\boldsymbol{u} \end{bmatrix}$$

where $\boldsymbol{u}$ is an n-dimensional vector, then

$$\boldsymbol{t}'\boldsymbol{\xi} = \boldsymbol{u}'\boldsymbol{z}$$

$$\boldsymbol{t}'\boldsymbol{M}\boldsymbol{t} = 0$$

$$\boldsymbol{t}'\boldsymbol{\gamma} = \boldsymbol{u}'\boldsymbol{c}$$

(where $\boldsymbol{z}$ and $\boldsymbol{c}$ are as defined in Section 1) and (2.3) becomes

$$\mathscr{E}e^{\boldsymbol{u}'\boldsymbol{z}} = e^{\boldsymbol{u}'\boldsymbol{c}} \quad . \tag{2.5}$$

Let

$$\boldsymbol{J} = \begin{bmatrix} \boldsymbol{I} & \boldsymbol{I} \\ i\boldsymbol{I} & -i\boldsymbol{I} \end{bmatrix}$$

where $\boldsymbol{I}$ is the $n \times n$ identity matrix. If we let

$$\boldsymbol{t} = \boldsymbol{J}\boldsymbol{v},$$

where $\boldsymbol{v}$ is a $2n$-dimensional vector, then

$$\boldsymbol{t}'\boldsymbol{\xi} = \boldsymbol{v}'\boldsymbol{\eta}$$

$$\boldsymbol{t}'\boldsymbol{M}\boldsymbol{t} = \boldsymbol{v}'\boldsymbol{\Sigma}\boldsymbol{v}$$

$$\boldsymbol{t}'\boldsymbol{\gamma} = \boldsymbol{v}'\boldsymbol{h}$$

(where $\boldsymbol{\eta}$, $\boldsymbol{h}$, and $\boldsymbol{\Sigma}$ are as defined in Section 1) and (2.3) becomes

$$\mathscr{E}\exp(\boldsymbol{v}'\boldsymbol{\eta}) = \exp(\tfrac{1}{2}\boldsymbol{v}'\boldsymbol{\Sigma}\boldsymbol{v} + \boldsymbol{v}'\boldsymbol{h}) \quad . \tag{2.6}$$

Equations (2.3), (2.5), and (2.6) enable us to compute the integral moments of the components of $\boldsymbol{z}' = \{z_1, \cdots, z_n\}$. For example, from (2.3)

$$\mathscr{E}x_1^{\alpha_1} y_1^{\beta_1} \cdots x_n^{\alpha_n} y_n^{\beta_n} = \frac{\partial^{\alpha_1+\cdots+\alpha_n+\beta_1+\cdots+\beta_n}}{\partial t_1^{\alpha_1}\cdots\partial t_n^{\alpha_n}\,\partial t_{n+1}^{\beta_1}\cdots\partial t_{2n}^{\beta_n}} \exp\left(\tfrac{1}{2}\boldsymbol{t}'\boldsymbol{M}\boldsymbol{t} + \boldsymbol{t}'\boldsymbol{\gamma}\right)\Bigg|_{\boldsymbol{t}=\boldsymbol{o}} \tag{2.7}$$

where $z_j = x_j + iy_j$, $1 \leqq j \leqq n$, and the α_k and β_l are nonnegative integers. From (2.5)

$$\mathscr{E} z_1^{\gamma_1} \cdots z_n^{\gamma_n} = \frac{\partial^{\gamma_1 + \cdots + \gamma_n}}{\partial u_1^{\gamma_1} \cdots \partial u_n^{\gamma_n}} e^{\mathbf{u}'\mathbf{c}} \bigg|_{\mathbf{u}=\mathbf{0}} = c_1^{\gamma_1} \cdots c_n^{\gamma_n} \tag{2.8}$$

where the γ_j, $1 \leqq j \leqq n$, are nonnegative integers and $\mathbf{c}' = \{c_1, \cdots, c_n\}$. If $c_k = 0$ for some k, $1 \leqq k \leqq n$, then in the formula above we replace $c_k^{\gamma_k}$ by one for $\gamma_k = 0$, and we replace $c_k^{\gamma_k}$ by zero for $\gamma_k > 0$. From (2.6)

$$\mathscr{E} z_1^{\delta_1} \bar{z}_1^{\varepsilon_1} \cdots z_n^{\delta_n} \bar{z}_n^{\varepsilon_n} = \frac{\partial^{\delta_1 + \cdots + \delta_n + \varepsilon_1 + \cdots + \varepsilon_n}}{\partial v_1^{\delta_1} \cdots \partial v_n^{\delta_n} \, \partial v_{n+1}^{\varepsilon_1} \cdots \partial v_{2n}^{\varepsilon_n}} \exp(\tfrac{1}{2}\mathbf{v}'\boldsymbol{\Sigma}\mathbf{v} + \mathbf{v}'\mathbf{h}) \bigg|_{\mathbf{v}=\mathbf{0}} \tag{2.9}$$

where the δ_k and ε_l are nonnegative integers.

Let us examine (2.7), (2.8), and (2.9) in more detail. From (2.8) we see that

$$\mathscr{E} \prod_{k=1}^{n} z_k^{\gamma_k} = \prod_{k=1}^{n} \mathscr{E} z_k^{\gamma_k}$$

and if z is a component of $\mathbf{z}$ with mean zero, $\mathscr{E} z = 0$, then $\mathscr{E} z^{\gamma} = 0$ for any positive integer γ. The other cases are more interesting. If we compare (2.3) with (3.7) of Chapter II, we see that their form is identical. Thus (2.7) is merely the formula for calculating the moments of the components of a real $2n$-dimensional Gaussian vector. However, the form of (2.6) is also identical with that of (3.7) of Chapter II. Thus we have the interesting, but not unexpected, conclusion that the formula (2.9) for the calculation of the moments of the complex components of a complex Gaussian vector is identical with that for the computation of the moments of the components of a real Gaussian vector.

For example, if z is a component of $\mathbf{z}$ with mean c and variance $\mathscr{E}|z - c|^2 = R > 0$, then

$$\mathscr{E}|z|^{2m} = m! R^m L_m(-|c|^2 R^{-1}) \tag{2.10}$$

where m is a nonnegative integer and $L_m(\cdot)$ is the Laguerre polynomial of degree m. If $c = 0$,

$$\mathscr{E}|z|^{2m} = m! R^m \quad . \tag{2.11}$$

Also, if z_i, z_j, z_k, z_l are (not necessarily distinct) components of $\mathbf{z}$, and $\mathscr{E}(\mathbf{z} - \mathbf{c})(\bar{\mathbf{z}} - \bar{\mathbf{c}})' = \mathbf{R} = \| R_{jk} \|$, then

$$\begin{aligned} \mathscr{E} z_i \bar{z}_j z_k \bar{z}_l = {} & R_{ij} R_{kl} + R_{il} R_{kj} + R_{ij} c_k \bar{c}_l + R_{il} \bar{c}_j c_k \\ & + R_{kj} c_i \bar{c}_l + R_{kl} c_i \bar{c}_j + c_i \bar{c}_j c_k \bar{c}_l \end{aligned} \tag{2.12}$$

[compare with (2.1)]. If $i = j$ and $k = l$, then

$$\operatorname{Cov}(|z_j|^2, |z_k|^2) = |\mathscr{E} z_j \bar{z}_k|^2 - |c_j|^2 |c_k|^2 \quad . \tag{2.13}$$

Suppose now that z_j and z_k are components of $\boldsymbol{z}$ and $\boldsymbol{z}$ has mean zero. Then $\mathscr{E} z_j = 0 = \mathscr{E} z_k$. For nonnegative integers p and q with $p \geqq q$,

$$\mathscr{E} |z_j|^{2p} |z_k|^{2q} = p!q!\, R_{jj}^{p} R_{kk}^{q} \sum_{\alpha=0}^{q} \binom{p}{\alpha} \binom{q}{\alpha} \lambda_{jk}^{2\alpha} \tag{2.14}$$

where

$$\lambda_{jk}^2 = \frac{|R_{jk}|^2}{R_{jj} R_{kk}} \quad . \tag{2.15}$$

In terms of the Jacobi polynomial $P_n^{(\alpha,\beta)}(\,\cdot\,)$ (see [AS, page 775]) we may write, for $p \geqq q$,

$$\mathscr{E} |z_j|^{2p} |z_k|^{2q} = p!q!\, R_{jj}^{p-q} (|R_{jk}|^2 - R_{jj} R_{kk})^q P_q^{(p-q,0)}(r_{jk}) \tag{2.16}$$

where

$$r_{jk} = \frac{|R_{jk}|^2 + R_{jj} R_{kk}}{|R_{jk}|^2 - R_{jj} R_{kk}} \quad .$$

3. Distributions of Amplitudes and Phases

Let $\boldsymbol{z}' = \{z_1, \cdots, z_n\}$ have a joint n-dimensional complex Gaussian distribution. We have written $\boldsymbol{z} = \boldsymbol{x} + i\boldsymbol{y}$ and defined the $2n$-dimensional real Gaussian vector $\boldsymbol{\xi}$ as

$$\boldsymbol{\xi} = \begin{bmatrix} \boldsymbol{x} \\ \boldsymbol{y} \end{bmatrix}$$

[see (1.3b)]. The frequency function of $\boldsymbol{\xi}$ is given by (1.7). If $\boldsymbol{z}$ has mean vector $\boldsymbol{c}$ and positive definite covariance matrix $\boldsymbol{R}$, then the marginal density function of $\boldsymbol{x}$ is n-dimensional Gaussian with mean $\boldsymbol{a} = \operatorname{Re} \boldsymbol{c}$ and covariance matrix $\frac{1}{2} \operatorname{Re} \boldsymbol{R}$. Similarly, $\boldsymbol{y}$ is normally distributed in n dimensions with mean vector $\boldsymbol{b} = \operatorname{Im} \boldsymbol{c}$ and the same covariance, $\frac{1}{2} \operatorname{Re} \boldsymbol{R}$.

Now instead of decomposing $\boldsymbol{z}$ into *Cartesian coordinates*, we may write it in terms of *polar coordinates*. Thus if we let

$$z_k = r_k \exp(i\theta_k) \quad , \qquad 1 \leqq k \leqq n \quad ,$$

where $r_k = |z_k|$, then we may consider the joint distribution of $\boldsymbol{r}' = \{r_1, \cdots, r_n\}$ and $\boldsymbol{\theta}' = \{\theta_1, \cdots, \theta_n\}$ as well as the marginal density functions of $\boldsymbol{r}$ and $\boldsymbol{\theta}$. The joint density function $f^*(\boldsymbol{r}, \boldsymbol{\theta})$ of $\boldsymbol{r}$ and $\boldsymbol{\theta}$ is related to the density function $f(\mathbf{z})$ of $\mathbf{z}$ [see (1.1), or (1.6), or (1.7)] by the equation

$$f^*(\mathbf{r}, \boldsymbol{\theta}) = \left[\prod_{k=1}^{n} r_k \right] f(\mathbf{z}) \quad . \tag{3.1}$$

Hence the marginal density function of $\mathbf{r}$,

$$g(\mathbf{r}) = \int_0^{2\pi} f^*(\mathbf{r}, \boldsymbol{\theta})\, d\boldsymbol{\theta}, \tag{3.2}$$

is a Rayleigh frequency function, and the marginal density function

$$h(\boldsymbol{\theta}) = \int_0^{\infty} f^*(\mathbf{r}, \boldsymbol{\theta})\, d\mathbf{r} \tag{3.3}$$

is a distribution of phases.

Unfortunately one cannot compute $g(\mathbf{r})$ and $h(\boldsymbol{\theta})$ in closed form for arbitrary n. We shall consider certain special cases and related formulas. (For further results of this type we refer the reader to [Mid], [Mil 1, 2, 3, 4], [MS 1, 2].)

Suppose $\mathbf{z}' = \{z\}$ is a one-dimensional complex Gaussian vector with mean c and variance R. Then we may write

$$z = re^{i\theta}$$

where $r = |z|$, and

$$\mathscr{E}z = c = |c| \exp(i\gamma)$$

$$\operatorname{Var} z = \mathscr{E}|z - c|^2 = R > 0 \quad .$$

From (1.1) for $n = 1$,

$$f(z) = \frac{1}{\pi R} \exp\left(\frac{-|z - c|^2}{R}\right)$$

and from (3.1)

$$f^*(r, \theta) = \frac{r}{\pi R} \exp\left(\frac{-(r^2 + |c|^2)}{R}\right) \exp\left[2|c| r R^{-1} \cos(\theta - \gamma)\right] \quad . \tag{3.4}$$

Therefore

$$\begin{aligned} g(r) &= \frac{r}{\pi R} \exp\left(\frac{-(r^2 + |c|^2)}{R}\right) \int_0^{2\pi} \exp\left[2|c| r R^{-1} \cos(\theta - \gamma)\right] d\theta \\ &= \frac{2r}{R} \exp\left(\frac{-(r^2 + |c|^2)}{R}\right) I_0\left(\frac{2|c|r}{R}\right) \quad , \qquad r \geqq 0 \\ &= 0 \quad , \qquad r < 0 \quad , \end{aligned} \tag{3.5}$$

where I_0 is the modified Bessel function of the first kind and order zero. If z has mean zero, that is, if $\mathscr{E}z = c = 0$, then

$$
\begin{aligned}
g(r) &= \frac{2r}{R} \exp\left(\frac{-r^2}{R}\right) , \qquad r \geqq 0 , \\
&= 0, \qquad r < 0 .
\end{aligned} \tag{3.6}
$$

The density function of the phase is given by

$$
\begin{aligned}
h(\theta) &= \frac{1}{\pi R} \exp\left(\frac{-|c|^2}{R}\right) \int_0^\infty r \exp\left(\frac{-r^2}{R}\right) \exp\left[2|c| r R^{-1} \cos(\theta - \gamma)\right] dr \\
&= \frac{1}{2\pi} \exp\left(\frac{-|c|^2[1 + \sin^2(\theta - \gamma)]}{2R}\right) D_{-2}\left(-|c|\left(\frac{2}{R}\right)^{1/2} \cos(\theta - \gamma)\right) , \\
&\qquad 0 \leqq \theta < 2\pi , \\
&= 0 , \qquad \text{otherwise} ,
\end{aligned} \tag{3.7}
$$

where $D_{-\nu}$ is the parabolic cylinder function. If $c = 0$, then

$$
\begin{aligned}
h(\theta) &= \frac{1}{2\pi} , \qquad 0 \leqq \theta < 2\pi , \\
&= 0 , \qquad \text{otherwise} .
\end{aligned} \tag{3.8}
$$

It is interesting to observe that

$$
u = \tan\theta = \frac{\operatorname{Im} z}{\operatorname{Re} z}
$$

is just the ratio of two independent real Gaussian variates. The frequency function of u is

$$
k(u) = \frac{1}{2\pi(1 + u^2)} \exp\left(\frac{-|c|^2}{R}\right) \exp\left(\frac{\mu^2}{4}\right)[D_{-2}(\mu) + D_{-2}(-\mu)]
$$

where

$$
\mu = \frac{\sqrt{2}|c|(\cos\gamma + u \sin\gamma)}{[R(1 + u^2)]^{1/2}} = |c|\left(\frac{2}{R}\right)^{1/2} \cos(\theta - \gamma) .
$$

If $c = 0$, then $D_{-2}(0) = 1$ and u is Cauchy distributed,

$$
k(u) = \frac{1}{\pi(1 + u^2)} .
$$

We turn now to the case where $n = 2$. Then $\mathbf{z}' = \{z_1, z_2\}$ is a two-dimensional complex Gaussian vector. We shall write

$$z_1 = r_1 \exp(i\theta_1) \quad , \qquad z_2 = r_2 \exp(i\theta_2)$$

where $r_1 = |z_1|$ and $r_2 = |z_2|$; and we assume $\boldsymbol{z}$ has mean zero and positive definite covariance matrix $\boldsymbol{R}$. Our formulas are expressed more conveniently in terms of the inverse covariance matrix,

$$\boldsymbol{R}^{-1} = \boldsymbol{S} = \begin{bmatrix} S_{11} & S_{12} \\ \bar{S}_{12} & S_{22} \end{bmatrix} \quad . \tag{3.9}$$

Clearly S_{11} and S_{22} are real and positive. We shall write

$$S_{12} = |S_{12}| \exp(i\chi_{12}) \quad . \tag{3.10}$$

From (1.1) and (3.1), for $n = 2$,

$$f^*(r_1, r_2, \theta_1, \theta_2) = \frac{r_1 r_2 \det \boldsymbol{S}}{\pi^2} \exp(-\bar{\boldsymbol{z}}'\boldsymbol{S}\boldsymbol{z}) \quad .$$

Since

$$\bar{\boldsymbol{z}}'\boldsymbol{S}\boldsymbol{z} = S_{11}r_1^2 + S_{22}r_2^2 + 2|S_{12}|r_1 r_2 \cos(\theta_1 - \theta_2 - \chi_{12}) \quad ,$$

the joint distribution of amplitudes is given by

$$\begin{aligned} g(r_1, r_2) &= \frac{r_1 r_2 \det \boldsymbol{S}}{\pi^2} \exp[-(S_{11}r_1^2 + S_{22}r_2^2)] \\ &\quad \times \int_0^{2\pi} \int_0^{2\pi} \exp[-2|S_{12}|r_1 r_2 \cos(\theta_1 - \theta_2 - \chi_{12})] \, d\theta_1 d\theta_2 \\ &= 4r_1 r_2 (\det \boldsymbol{S}) \exp[-(S_{11}r_1^2 + S_{22}r_2^2)] \, I_0(2|S_{12}|r_1 r_2) \quad , \\ &\qquad\qquad r_1, r_2 \geqq 0 \quad , \\ &= 0, \qquad\qquad \text{otherwise} \quad , \end{aligned} \tag{3.11}$$

and the joint density function of phases is

$$\begin{aligned} h(\theta_1, \theta_2) &= \frac{\det \boldsymbol{S}}{\pi^2} \int_0^\infty \int_0^\infty r_1 r_2 \exp[-(S_{11}r_1^2 + S_{22}r_2^2)] \\ &\quad \times \exp[-2|S_{12}|r_1 r_2 \cos(\theta_1 - \theta_2 - \chi_{12})] \, dr_1 dr_2 \\ &= \frac{\det \boldsymbol{S}}{8\pi^2 S_{11} S_{22}} \frac{\partial^2}{\partial\lambda^2} (\text{Arccos}\,\lambda)^2 \quad , \qquad 0 \leqq \theta_1, \theta_2 < 2\pi \quad , \\ &= 0 \quad , \qquad\qquad \text{otherwise} \end{aligned} \tag{3.12}$$

where

$$\lambda = \frac{|S_{12}|}{(S_{11}S_{22})^{1/2}} \cos(\theta_1 - \theta_2 - \chi_{12}) \quad .$$

If we let

$$u_1 = \tan\theta_1 = \frac{\text{Im } z_1}{\text{Re } z_1}$$

and

$$u_2 = \tan\theta_2 = \frac{\text{Im } z_2}{\text{Re } z_2} ,$$

then the joint frequency function of u_1 and u_2 is

$$k(u_1, u_2) = \frac{\det \boldsymbol{S}}{2\pi^2(1 + u_1^2)(1 + u_2^2)\, S_{11}S_{22}} \frac{\partial^2}{\partial \nu^2} (\text{Arcsin } \nu)^2$$

where

$$\begin{aligned} \nu &= \frac{|S_{12}|[(1 + u_1u_2)\cos\chi_{12} + (u_1 - u_2)\sin\chi_{12}]}{\sqrt{(1 + u_1^2)(1 + u_2^2)\, S_{11}S_{22}}} \\ &= \frac{|S_{12}|}{(S_{11}S_{12})^{1/2}} \cos(\theta_1 - \theta_2 - \chi_{12}) \\ &= \lambda \end{aligned}$$

(and we see that $|\nu| < 1$ since $|\lambda| < 1$). If $S_{12} = 0$, then u_1 and u_2 are independent and Cauchy distributed,

$$k(u_1, u_2) = k(u_1)\,k(u_2) = \frac{1}{\pi^2(1 + u_1^2)(1 + u_2^2)} .$$

Let $\boldsymbol{z}_k' = \{z_{k,1},\, z_{k,2}\}$, $1 \leq k \leq m$, be m independent two-dimensional complex Gaussian vectors with means zero and identical positive definite covariance matrix $\boldsymbol{R}$ [see (3.9)], and consider

$$Z = \sum_{k=1}^{m} z_{k,1}\, \bar{z}_{k,2} .$$

Then the joint density function H of $u = \text{Re } Z$ and $\nu = \text{Im } Z$ is

$$\begin{aligned} H(u, \nu) = {} & \frac{2(\det \boldsymbol{S})^m (u^2 + \nu^2)^{(m-1)/2}}{\pi(m - 1)!(S_{11}S_{22})^{(m-1)/2}} \\ & \times \exp[-2|S_{12}|(u\cos\chi_{12} + \nu\sin\chi_{12})] \\ & \times K_{m-1}(2[S_{11}S_{22}(u^2 + \nu^2)]^{1/2}) \end{aligned}$$

where K_ν, is the modified Bessel function of the second kind and order ν. We deduce the equation above by showing that the joint characteristic function $\psi(t, s)$ of u and ν is the mth power of

$$\frac{4 \det \boldsymbol{S}}{t^2 + s^2 + 4i|S_{12}|(t \cos \chi_{12} + s \sin \chi_{12}) + 4 \det \boldsymbol{S}} \quad .$$

Finally, we shall compute $g(r_1, r_2, r_3)$ where $\boldsymbol{z}' = \{z_1, z_2, z_3\}$ is a three-dimensional complex Gaussian vector, $\mathscr{E}\boldsymbol{z} = \boldsymbol{0}$, the covariance matrix $\boldsymbol{R} = \mathscr{E}\boldsymbol{z}\bar{\boldsymbol{z}}'$ is positive definite, and

$$z_k = r_k \exp(i\theta_k) \quad , \qquad k = 1, 2, 3 \quad .$$

As in the two-dimensional case, we prefer to work with the inverse covariance matrix $\boldsymbol{S}$,

$$\boldsymbol{R}^{-1} = \boldsymbol{S} = \begin{bmatrix} S_{11} & S_{12} & S_{13} \\ S_{21} & S_{22} & S_{23} \\ S_{31} & S_{32} & S_{33} \end{bmatrix} \quad , \qquad S_{jk} = \bar{S}_{kj} \quad .$$

We shall write

$$S_{12} = |S_{12}| \exp(i\chi_{12}) \quad , \qquad S_{23} = |S_{23}| \exp(i\chi_{23}) \quad ,$$
$$S_{31} = |S_{31}| \exp(i\chi_{31}) \quad .$$

The positive definite quadratic form $\bar{\boldsymbol{z}}'\boldsymbol{S}\boldsymbol{z}$ may be partially expanded as

$$\bar{\boldsymbol{z}}'\boldsymbol{S}\boldsymbol{z} = A(\theta_1, \theta_3) \cos \theta_2 + B(\theta_1, \theta_3) \sin \theta_2 + C(\theta_3, \theta_1) + r_1^2 S_{11} + r_2^2 S_{22} + r_3^2 S_{33}$$

where A, B, and C do not depend on θ_2. Thus, again invoking (3.2)

$$g(r_1, r_2, r_3) = \frac{\det \boldsymbol{S}}{\pi^3} r_1 r_2 r_3 \exp[-(r_1^2 S_{11} + r_2^2 S_{22} + r_3^2 S_{33})]$$
$$\times \int_0^{2\pi} \int_0^{2\pi} \int_0^{2\pi} \exp[-(A \cos \theta_2 + B \sin \theta_2 + C)] \, d\theta_1 \, d\theta_2 \, d\theta_3 \quad . \quad (3.13)$$

But

$$\int_0^{2\pi} \exp[-(A \cos \theta_2 + B \sin \theta_2)] \, d\theta_2 = 2\pi I_0((A^2 + B^2)^{1/2})$$

and

$$A^2 + B^2 = 4r_2^2 [r_1^2 |S_{12}|^2 + r_3^2 |S_{23}|^2 - 2r_1 r_3 |S_{12}| \, |S_{23}| \times \cos(\theta_1 - \theta_3 - \chi_{12} - \chi_{23} + \pi)] \quad .$$

From the Neumann addition formula we have

$$I_0((A^2 + B^2)^{1/2}) = \sum_{m=0}^{\infty} \varepsilon_m I_m(2r_1r_2|S_{12}|)\, I_m(2r_2r_3|S_{23}|)$$
$$\times \cos m(\theta_1 - \theta_3 - \chi_{12} - \chi_{23})$$

where ε_m is the Neumann factor ($\varepsilon_0 = 1$, $\varepsilon_m = 2$ for $m = 1, 2, \cdots$) and I_m is the modified Bessel function of the first kind and order m. Since

$$C(\theta_3, \theta_1) = 2r_3r_1|S_{31}|\cos(\theta_3 - \theta_1 - \chi_{31})$$

we have

$$\int_0^{2\pi}\int_0^{2\pi} \exp[-C(\theta_3, \theta_1)] \cos m(\theta_1 - \theta_3 - \chi_{12} - \chi_{23})\, d\theta_1\, d\theta_3$$
$$= 4\pi^2(-1)^m I_m(2r_3r_1|S_{31}|) \cos m(\chi_{12} + \chi_{23} + \chi_{31}) \qquad (3.14)$$

and hence (3.13) reduces to

$$g(r_1, r_2, r_3) = 8(\det \boldsymbol{S})\, r_1r_2r_3 \exp[-(r_1^2S_{11} + r_2^2S_{22} + r_3^2S_{33})]$$
$$\times \sum_{m=0}^{\infty} \varepsilon_m(-1)^m I_m(2r_1r_2|S_{12}|)\, I_m(2r_2r_3|S_{23}|)$$
$$\times I_m(2r_3r_1|S_{31}|) \cos m(\chi_{12} + \chi_{23} + \chi_{31}) \ . \qquad (3.15)$$

4. Some Nonintegral Moments and Moments of the Phases

The availability of the density functions of the amplitudes and phases of $\boldsymbol{z}$ allows us to compute nonintegral moments of the amplitudes and moments of the phases. For example, if Re $\alpha > -2$, then from (3.5)

$$\mathscr{E}r^\alpha = \int_0^\infty r^\alpha g(r)\, dr$$
$$= \frac{2}{R} \exp\left(\frac{-|c|^2}{R}\right) \int_0^\infty r^{1+\alpha} \exp\left(\frac{-r^2}{R}\right) I_0(2|c|rR^{-1})\, dr \ .$$

If we expand the Bessel function and integrate term by term, we obtain

$$\mathscr{E}r^\alpha = \Gamma(1 + \tfrac{1}{2}\alpha)\, R^{\alpha/2} \exp\left(\frac{-|c|^2}{R}\right) {}_1F_1(1 + \tfrac{1}{2}\alpha, 1; |c|^2R^{-1}) \qquad (4.1)$$

where ${}_1F_1$ is the confluent hypergeometric function. If $\mathscr{E}z = c = 0$, then

$$\mathscr{E}r^\alpha = \Gamma(1 + \tfrac{1}{2}\alpha)R^{\alpha/2} \ , \qquad \text{Re}\, \alpha > -2 \ , \qquad (4.2)$$

and if, in addition, α is an even integer, say, $\alpha = 2m$, then

$$r^\alpha = r^{2m} = |z|^{2m}$$

and

$$\mathscr{E}r^{\alpha} = \mathscr{E}|z|^{2m} = m!R^m \quad , \tag{4.3}$$

[compare with (2.11)].

If $\alpha = -1$, then from (4.1),

$$\mathscr{E}\frac{1}{r} = \Gamma(\tfrac{1}{2})R^{-1/2} \exp\left(\frac{-|c|^2}{R}\right) {}_1F_1(\tfrac{1}{2}, 1; |c|^2 R^{-1})$$

$$= \left(\frac{\pi}{R}\right)^{1/2} \exp\left(\frac{-|c|^2}{2R}\right) I_0(\tfrac{1}{2}|c|^2 R^{-1}) \quad .$$

Thus we have found $\mathscr{E}(1/|z|)$ (since $r = |z|$). It is interesting to compare this with $\mathscr{E}(1/z)$.

To calculate $\mathscr{E}(1/z)$ we have, from (1.1) with $n = 1$, that

$$\mathscr{E}\frac{1}{z} = \frac{1}{\pi R}\int_{-\infty}^{\infty}\int_{-\infty}^{\infty} \frac{1}{z} \exp\left(\frac{-|z-c|^2}{R}\right) dx\, dy$$

where we have written $z = x + iy$. A change to polar coordinates,

$$x = r\cos\theta$$

$$y = r\sin\theta$$

and an integration with respect to θ leads to

$$\mathscr{E}\frac{1}{z} = \frac{2}{R}\exp(-i\gamma)\exp\left(\frac{-|c|^2}{R}\right)\int_0^{\infty}\exp\left(\frac{-r^2}{R}\right) I_1(2|c|rR^{-1})\, dr$$

where I_1 is the modified Bessel function of the first kind and order one. Continuing,

$$\mathscr{E}\frac{1}{z} = \frac{|c|}{R}\exp(-i\gamma)\exp\left(\frac{-|c|^2}{R}\right) {}_1F_1(1, 2; |c|^2 R^{-1}) \quad .$$

But

$${}_1F_1(1, 2; \xi) = \frac{1}{\xi}(e^{\xi} - 1) \quad .$$

Hence

$$\mathscr{E}\frac{1}{z} = \frac{1}{c}\left[1 - \exp\left(\frac{-|c|^2}{R}\right)\right] \quad .$$

The computation of $\mathscr{E}\theta$ is more complicated. From (3.7)

$$\mathscr{E}\theta = \int_0^{2\pi} \theta h(\theta)\, d\theta$$
$$= \frac{1}{\pi R} \exp\left(\frac{-|c|^2}{R}\right) \int_0^{\infty} r \exp\left(\frac{-r^2}{R}\right) dr$$
$$\times \int_0^{2\pi} \theta \exp\left[2|c|rR^{-1} \cos(\theta - \gamma)\right] d\theta \quad . \tag{4.4}$$

We may use the Jacobi–Anger formula

$$e^{z \cos \psi} = \sum_{m=0}^{\infty} \varepsilon_m I_m(z) \cos m\psi \tag{4.5}$$

to conclude that

$$\int_0^{2\pi} \theta \exp[2|c|rR^{-1} \cos(\theta - \gamma)]\, d\theta$$
$$= 2\pi^2 I_0(2|c|rR^{-1}) - 4\pi \sum_{m=1}^{\infty} \frac{\sin m\gamma}{m} I_m(2|c|rR^{-1}) \tag{4.6}$$

since

$$\int_0^{2\pi} \theta \cos m(\theta - \gamma)\, d\theta = -\frac{2\pi}{m} \sin m\gamma \quad , \qquad m > 0 \quad .$$

Now substitute (4.6) in (4.4) and recall that

$$\int_0^{\infty} x \exp(-\alpha x^2)\, I_\nu(2\beta x)\, dx = \frac{\beta^\nu}{2\alpha^{1+(\nu/2)}} \frac{\Gamma(1 + \frac{1}{2}\nu)}{\Gamma(1 + \nu)} \quad .$$
$$\times\ {}_1F_1(1 + \tfrac{1}{2}\nu, \nu + 1;\ \beta^2\alpha^{-1}) \tag{4.7}$$

for $\alpha > 0$ and Re $\nu > -2$. Then

$$\mathscr{E}\theta = \pi - 2 \exp\left(\frac{-|c|^2}{R}\right) \sum_{m=1}^{\infty} \frac{\Gamma(1 + \frac{1}{2}m)}{m\Gamma(1 + m)} \left(\frac{|c|}{\sqrt{R}}\right)^m$$
$$\times \sin m\gamma\ {}_1F_1(1 + \tfrac{1}{2}m, 1 + m; |c|^2 R^{-1}) \quad . \tag{4.8}$$

The identity

$$e^{-x}\ {}_1F_1(\alpha, \beta;\ x) = {}_1F_1(\beta - \alpha, \beta;\ -x) \tag{4.9}$$

yields the alternate form

$$\mathscr{E}\theta = \pi - 2 \sum_{m=1}^{\infty} \frac{\Gamma(1 + \frac{1}{2}m)}{m\Gamma(1 + m)} \left(\frac{|c|}{\sqrt{R}}\right)^m$$
$$\times \sin m\gamma\ {}_1F_1(\tfrac{1}{2}m, 1 + m;\ -|c|^2 R^{-1}) \tag{4.10}$$

for $\mathscr{E}\theta$. Note that if γ is an integral multiple of π, that is, if the mean of z is *real*, then

$$\mathscr{E}\theta = \pi \quad . \tag{4.11}$$

If m is an integer, then we may also compute $\mathscr{E}\exp(im\theta)$ from (3.7), namely

$$\begin{aligned}\mathscr{E}\exp(im\theta) &= \frac{\Gamma(1+\frac{1}{2}m)}{\Gamma(1+m)}\frac{|c|^m}{R^{m/2}}\exp(im\gamma)\exp\left(\frac{-|c|^2}{R}\right) \\ &\quad \times {}_1F_1(1+\tfrac{1}{2}m, 1+m; |c|^2R^{-1}) \qquad (4.12) \\ &= \frac{\Gamma(1+\frac{1}{2}m)}{\Gamma(1+m)}\frac{|c|^m}{R^{m/2}}\exp(im\gamma)\,{}_1F_1(\tfrac{1}{2}m, 1+m; -|c|^2R^{-1})\end{aligned}$$

where we have used (4.9). If $c = 0$, then for arbitrary real β,

$$\begin{aligned}\mathscr{E}e^{i\beta\theta} &= e^{i\pi\beta}\left(\frac{\sin\pi\beta}{\pi\beta}\right) \quad , \qquad \beta \neq 0 \quad , \\ &= 1 \quad , \qquad\qquad\qquad \beta = 0 \quad .\end{aligned} \tag{4.13}$$

Thus if β is a nonzero integer, the expected value of $e^{i\beta\theta}$ is zero.

Now let us consider the two-dimensional case. Then from (3.11) for Re $\alpha > -2$, Re $\beta > -2$

$$\mathscr{E}r_1^\alpha r_2^\beta = \int_0^\infty\int_0^\infty r_1^\alpha r_2^\beta g(r_1, r_2)\,dr_1\,dr_2 \quad .$$

If we expand the Bessel function in g and integrate term by term, there results

$$\mathscr{E}r_1^\alpha r_2^\beta = \frac{\Gamma(1+\frac{1}{2}\alpha)\,\Gamma(1+\frac{1}{2}\beta)(\det \boldsymbol{S})}{S_{11}^{1+\frac{1}{2}\alpha}\,S_{22}^{1+\frac{1}{2}\beta}} \times {}_2F_1(1+\tfrac{1}{2}\alpha, 1+\tfrac{1}{2}\beta, 1; \lambda_{12}^2) \tag{4.14}$$

where ${}_2F_1$ is the hypergeometric function and

$$\lambda_{12}^2 = \frac{|S_{12}|^2}{S_{11}S_{22}} \quad . \tag{4.15}$$

If α and β are even integers, say $\alpha = 2p$, $\beta = 2q$, then

$$r_1^\alpha r_2^\beta = |z_1|^{2p}|z_2|^{2q}$$

and from (4.14)

$$\mathscr{E}|z_1|^{2p}|z_2|^{2q} = \frac{p!q!(\det \boldsymbol{S})}{S_{11}^{1+p}\,S_{22}^{1+q}}\,{}_2F_1(1+p, 1+q, 1; \lambda_{12}^2) \quad . \tag{4.16}$$

Let us show that this reduces to (2.14). First,

$$
\begin{aligned}
{}_2F_1(1+p, 1+q, 1; x) &= \frac{1}{p!q!}\frac{d^q}{dx^q}\left[x^q \frac{d^p}{dx^p}\left(\frac{x^p}{1-x}\right)\right] \\
&= \frac{1}{q!}\frac{d^q}{dx^q}[x^q(1-x)^{-p-1}] \\
&= \sum_{k=0}^{q}\binom{q}{k}\binom{p+k}{k}x^k(1-x)^{-(p+k+1)} \quad .
\end{aligned}
$$

Now

$$
(1-x)^\alpha = \sum_{k=0}^{\alpha}\binom{\alpha}{k}x^k(-1)^k \quad .
$$

If we multiply by x^p, differentiate p times with respect to x and then set x equal to one, we obtain

$$
\sum_{k=0}^{\alpha}\binom{\alpha}{k}\binom{p+k}{k}(-1)^{\alpha-k} = \binom{p}{\alpha} \quad .
$$

But

$$
\begin{aligned}
\sum_{k=0}^{q}\binom{q}{k}\binom{p+k}{k}x^k(1-x)^{q-k} &= \sum_{\alpha=0}^{q}\binom{q}{\alpha}\left[\sum_{k=0}^{\alpha}\binom{\alpha}{k}\binom{p+k}{k}(-1)^{\alpha-k}\right]x^\alpha \\
&= \sum_{\alpha=0}^{q}\binom{q}{\alpha}\binom{p}{\alpha}x^\alpha \quad .
\end{aligned}
$$

Thus

$$
{}_2F_1(1+p, 1+q, 1; x) = (1-x)^{-(p+q+1)}\sum_{\alpha=0}^{q}\binom{p}{\alpha}\binom{q}{\alpha}x^\alpha \quad ,
$$

and (4.16) becomes

$$
\mathscr{E}|z_1|^{2p}|z_2|^{2q} = \frac{p!q!(\det \boldsymbol{S})}{S_{11}^{1+p}\,S_{22}^{1+q}}(1-\lambda_{12}^2)^{-(p+q+1)}\sum_{\alpha=0}^{q}\binom{p}{\alpha}\binom{q}{\alpha}\lambda_{12}^{2\alpha} \quad .
$$

But if we write $\boldsymbol{R} = \boldsymbol{S}^{-1} = \| R_{jk} \|$, then

$$
\lambda_{12}^2 = \frac{|S_{12}|^2}{S_{11}S_{22}} = \frac{|R_{12}|^2}{R_{11}R_{22}} \tag{4.17}
$$

[see (4.15)] and

$$
\mathscr{E}|z_1|^{2p}|z_2|^{2q} = p!q!\; R_{11}^p\, R_{22}^q \sum_{\alpha=0}^{q}\binom{p}{\alpha}\binom{q}{\alpha}\lambda_{12}^{2\alpha} \quad ,
$$

which is (2.14).

Another special case of (4.14) leads to the covariance function of r_1 and r_2, namely,

$$\begin{aligned} \operatorname{Cov}(r_1, r_2) &= \mathscr{E} r_1 r_2 - \mathscr{E} r_1 \mathscr{E} r_2 \\ &= \mathscr{E} r_1 r_2 - \frac{\pi}{4} (R_{11} R_{22})^{1/2} \end{aligned} \tag{4.18}$$

by (4.2). But from (4.14)

$$\mathscr{E} r_1 r_2 = \frac{\pi}{4(S_{11} S_{22})^{1/2}} (1 - \lambda_{12}^2)\, {}_2F_1(\tfrac{3}{2}, \tfrac{3}{2}, 1; \lambda_{12}^2) \quad . \tag{4.19}$$

This formula may be further simplified if we recall that

$$\begin{aligned} (1 - \lambda^2)^2\, {}_2F_1(\tfrac{3}{2}, \tfrac{3}{2}, 1; \lambda^2) &= {}_2F_1(-\tfrac{1}{2}, -\tfrac{1}{2}, 1; \lambda^2) \\ &= \frac{2}{\pi} [2E(\lambda) - (1 - \lambda^2) K(\lambda)] \end{aligned}$$

where K and E are the complete elliptic integrals of the first and second kind, respectively. Specifically,

$$\mathscr{E} r_1 r_2 = \tfrac{1}{2} (R_{11} R_{22})^{1/2} [2E(\lambda_{12}) - (1 - \lambda_{12}^2) K(\lambda_{12})] \quad .$$

As in the one-dimensional case we may compute $\mathscr{E}(r_1/r_2)$ and $\mathscr{E}(1/r_1 r_2)$ from (4.14). Thus, with $\alpha = 1$, $\beta = -1$,

$$\mathscr{E} \frac{r_1}{r_2} = \frac{\pi \det \boldsymbol{S}}{2 S_{11}^{3/2} S_{22}^{1/2}}\, {}_2F_1(\tfrac{3}{2}, \tfrac{1}{2}, 1; \lambda_{12}^2) \quad .$$

If we use the identity

$${}_2F_1(a, b, c; x) = (1 - x)^{c-a-b}\, {}_2F_1(c - a, c - b, c; x) \tag{4.20}$$

with $a = \frac{3}{2}$, $b = \frac{1}{2}$, $c = 1$, then we may write

$$\begin{aligned} \mathscr{E} \frac{r_1}{r_2} &= \left(\frac{S_{22}}{S_{11}}\right)^{1/2} E(\lambda_{12}) \\ &= \left(\frac{R_{11}}{R_{22}}\right)^{1/2} E(\lambda_{12}) \end{aligned}$$

since

$$E(\lambda) = \tfrac{1}{2}\pi\, {}_2F_1(-\tfrac{1}{2}, \tfrac{1}{2}, 1; \lambda^2) \quad . \tag{4.21}$$

Also, with $\alpha = \beta = -1$, we obtain

$$\mathscr{E} \frac{1}{r_1 r_2} = \frac{\pi \det \boldsymbol{S}}{(S_{11} S_{22})^{1/2}}\, {}_2F_1(\tfrac{1}{2}, \tfrac{1}{2}, 1; \lambda_{12}^2)$$

$$= \frac{2 \det \boldsymbol{S}}{(S_{11}S_{22})^{1/2}} K(\lambda_{12})$$

$$= \frac{2}{(R_{11}R_{22})^{1/2}} K(\lambda_{12})$$

since

$$K(\lambda) = \tfrac{1}{2}\pi \; {}_2F_1(\tfrac{1}{2}, \tfrac{1}{2}, 1; \lambda^2) \quad . \tag{4.22}$$

Let us now calculate $\mathscr{E}(z_1/z_2)$ and $\mathscr{E}(1/\bar{z}_1 z_2)$. We have from (1.1) with $n = 2$ that

$$\mathscr{E}\frac{z_1}{z_2} = -\frac{4 \exp(i\chi_{12})}{\det \boldsymbol{R}} \int_0^\infty \int_0^\infty r_1^2 \exp[-(S_{11}r_1^2 + S_{22}r_2^2)]$$
$$\times I_1(2|S_{12}|r_1 r_2) \, dr_1 \, dr_2$$

and

$$\mathscr{E}\frac{1}{\bar{z}_1 z_2} = -\frac{4 \exp(i\chi_{12})}{\det \boldsymbol{R}} \int_0^\infty \int_0^\infty \exp[-(S_{11}r_1^2 + S_{22}r_2^2)]$$
$$\times I_1(2|S_{12}|r_1 r_2) \, dr_1 \, dr_2$$

where we have made the change of variables

$$z_1 = x_1 + iy_1 = r_1 \exp(i\theta_1)$$

$$z_2 = x_2 + iy_2 = r_2 \exp(i\theta_2)$$

to polar coordinates and used (3.14). The same type of argument that led to (4.14) now yields

$$\mathscr{E}\frac{z_1}{z_2} = -\frac{|S_{12}|\exp(i\chi_{12}) \det \boldsymbol{S}}{S_{11}^2 \, S_{22}} \, {}_2F_1(2, 1, 2; \lambda_{12}^2)$$

$$= -\frac{S_{12} \det \boldsymbol{S}}{S_{11}^2 S_{22}} \, {}_1F_0(1; \lambda_{12}^2)$$

$$= -\frac{S_{12}}{S_{11}} = \frac{R_{12}}{R_{22}}$$

and

$$\mathscr{E}\frac{1}{\bar{z}_1 z_2} = -\frac{|S_{12}|\exp(i\chi_{12}) \det \boldsymbol{S}}{S_{11}S_{22}} \, {}_2F_1(1, 1, 2; \lambda_{12}^2)$$

$$= (\lambda_{12}^{-2} - 1) S_{12} \log(1 - \lambda_{12}^2)$$

$$= -\bar{R}_{12}^{-1} \log(1 - \lambda_{12}^2) \quad .$$

For further formulas in this direction see [McG].

We now turn to the computation of the covariance function $\mathrm{Cov}(\theta_1, \theta_2)$ of the phases, or equivalently

$$\begin{aligned}\mathscr{E}\theta_1\theta_2 &= \mathrm{Cov}(\theta_1, \theta_2) + \mathscr{E}\theta_1\mathscr{E}\theta_2 \\ &= \mathrm{Cov}\,(\theta_1, \theta_2) + \pi^2\end{aligned} \tag{4.23}$$

by (4.11). From (3.12)

$$\mathscr{E}\theta_1\theta_2 = \int_0^{2\pi}\int_0^{2\pi} \theta_1\theta_2 h(\theta_1, \theta_2)\, d\theta_1\, d\theta_2 \quad .$$

Now we proceed as in the computation of $\mathscr{E}\theta$ [see (4.10)]. Use the Jacobi–Anger formula [see (4.5)] to write

$$\begin{aligned}&\int_0^{2\pi}\int_0^{2\pi} \theta_1\theta_2 \exp\left[-2|S_{12}|r_1r_2\cos(\theta_1 - \theta_2 - \chi_{12})\right] d\theta_1\, d\theta_2 \\ &\quad = 4\pi^4 I_0(2|S_{12}|r_1r_2) + 8\pi^2 \sum_{m=1}^{\infty} \frac{(-1)^m}{m^2} I_m(2|S_{12}|r_1r_2)\cos m\chi_{12} \quad .\end{aligned}$$

Then

$$\begin{aligned}&\int_0^{\infty}\int_0^{\infty} r_1r_2 \exp\left[-(S_{11}r_1^2 + S_{22}r_2^2)\right] I_m(2|S_{12}|r_1r_2)\, dr_1\, dr_2 \\ &\quad = \frac{\lambda_{12}^m}{4S_{11}S_{22}} \frac{\Gamma^2(1 + \frac{1}{2}m)}{\Gamma(1 + m)}\, {}_2F_1(1 + \tfrac{1}{2}m, 1 + \tfrac{1}{2}m, 1 + m; \lambda_{12}^2) \quad ,\end{aligned}$$

where λ_{12} is given by (4.17), is deduced by expanding the Bessel function and integrating term by term. Combining these results leads to

$$\begin{aligned}\mathscr{E}\theta_1\theta_2 &= \pi^2 + 2(1 - \lambda_{12}^2)\sum_{m=1}^{\infty} \frac{(-1)^m}{m^2}\lambda_{12}^m \frac{\Gamma^2(1 + \frac{1}{2}m)}{\Gamma(1 + m)}\cos m\chi_{12} \\ &\quad \times {}_2F_1(1 + \tfrac{1}{2}m, 1 + \tfrac{1}{2}m, 1 + m; \lambda_{12}^2) \quad .\end{aligned} \tag{4.24}$$

Now if we write $R_{12} = |R_{12}|e^{i\phi_{12}}$, then $\phi_{12} - \chi_{12}$ is congruent to π modulo 2π, and we may replace $(-1)^m \cos m\chi_{12}$ in (4.21) by $\cos m\phi_{12}$. Furthermore, if we use the identity (4.20) with $a = b = 1 + \frac{1}{2}m$ and $c = 1 + m$, then [see (4.23)] we have

$$\begin{aligned}\mathrm{Cov}(\theta_1, \theta_2) &= 2\sum_{m=1}^{\infty} \frac{1}{m^2}\lambda_{12}^m \frac{\Gamma^2(1 + \frac{1}{2}m)}{\Gamma(1 + m)}\cos m\phi_{12} \\ &\quad \times {}_2F_1(\tfrac{1}{2}m, \tfrac{1}{2}m, m + 1; \lambda_{12}^2) \quad .\end{aligned} \tag{4.25}$$

Let us now examine the joint moments of $e^{i\theta}$. If p and q are integers, then we have from (3.12), again, that

$$\mathscr{E} e^{ip\theta_1} e^{-iq\theta_2} = \delta_{pq} e^{ip\phi_{12}}(1 - \lambda_{12}^2) \lambda_{12}^p \frac{\Gamma^2(1 + \frac{1}{2}p)}{\Gamma(1 + p)} \times {}_2F_1(1 + \tfrac{1}{2}p, 1 + \tfrac{1}{2}p, 1 + p; \lambda_{12}^2) \tag{4.26}$$

where δ_{pq} is the Kronecker delta. In particular by (4.13)

$$\begin{aligned} \mathrm{Cov}(e^{i\theta_1}, \overline{e^{i\theta_2}}) &= \mathscr{E} e^{i\theta_1} e^{-i\theta_2} \\ &= \frac{\pi}{4} e^{i\phi_{12}}(1 - \lambda_{12}^2)\lambda_{12}\, {}_2F_1(\tfrac{3}{2}, \tfrac{3}{2}, 2; \lambda_{12}^2) \quad . \end{aligned} \tag{4.27}$$

This formula may also be expressed in terms of the complete elliptic integrals of the first and second kind. First we consider the identity

$$\begin{aligned} (\gamma - \alpha)\, x^{\gamma-\alpha-1}(1 - x)^{\alpha+\beta-\gamma-1}\, {}_2F_1(\alpha - 1, \beta, \gamma; x) \\ = \frac{d}{dx}[x^{\gamma-\alpha}(1 - x)^{\alpha+\beta-\gamma}\, {}_2F_1(\alpha, \beta, \gamma; x)] \quad . \end{aligned} \tag{4.28}$$

Then with $\alpha = \beta = \frac{1}{2}$ and $\gamma = 1$, it becomes (with $x = \lambda^2$)

$$2\lambda^2(1 - \lambda^2)\frac{dK(\lambda)}{d(\lambda^2)} = E(\lambda) - (1 - \lambda^2) K(\lambda) \quad , \tag{4.29}$$

see (4.21) and (4.22). The further formula

$$\frac{\alpha\beta}{\gamma}\, {}_2F_1(\alpha + 1, \beta + 1, \gamma + 1; x) = \frac{d}{dx}\, {}_2F_1(\alpha, \beta, \gamma; x) \tag{4.30}$$

with $\alpha = \beta = \frac{1}{2}$ and $\gamma = 1$ yields

$${}_2F_1(\tfrac{3}{2}, \tfrac{3}{2}, 2; \lambda^2) = \frac{8}{\pi}\frac{dK(\lambda)}{d(\lambda^2)} \quad . \tag{4.31}$$

Substitution of (4.31) and (4.29) into (4.27) leads to

$$\mathscr{E} e^{i\theta_1}\, \overline{e^{i\theta_2}} = \frac{1}{\lambda_{12}} e^{i\phi_{12}} [E(\lambda_{12}) - (1 - \lambda_{12}^2) K(\lambda_{12})] \quad . \tag{4.32}$$

From (4.25) and (4.26) we also have the interesting identity

$$\mathrm{Cov}(\theta_1, \theta_2) = 2 \sum_{m=1}^{\infty} \frac{1}{m^2} \cos m\phi_{12}\, \mathscr{E} \exp [im(\theta_1 - \theta_2 - \phi_{12})] \quad . \tag{4.33}$$

[We have again used (4.20).]

IV

STOCHASTIC DIFFERENTIAL EQUATIONS

0. Introduction

Let $\mathscr{L}$ be a linear differential operator, and let $\{\xi(t) \mid t \in T\}$ be a stochastic process. We wish to consider the process $\{z(t) \mid t \in T\}$ defined by the stochastic differential equation $\mathscr{L} z(t) = \xi(t)$. In particular we would like to relate the covariance function of the $\{z(t)\}$ process to the covariance function of the $\{\xi(t)\}$ process, and, when $\{\xi(t)\}$ is wide-sense stationary, relate the covariance function of the $\{z(t)\}$ process to the spectral distribution function of the $\{\xi(t)\}$ process. We accomplish these ends in Sections 3 and 5. Most of the difficulty in establishing our results stems from integrals: stochastic integrals, infinite integrals, Stieltjes integrals; and the problems of interchanging the order of integration.

In Section 1 we consider the stochastic process $\{\xi(t) \mid t \in [a, b]\}$ and investigate conditions under which the ubiquitous symbol $\int_a^b \xi(t)dt$ is meaningful as a sample function integral, as well as conditions under which it exists as a mean square integral. Various theorems sufficient to guarantee the existence and equality (with probability one) of these two integrals are also given. Generalizations to integrals of the form $\int_a^b g(t)\xi(t)dt$ and $\int_a^t K(t, s)\xi(s)ds$ where g and K are nonrandom functions are presented in Section 2, as well as theorems regarding the existence and equality of $\int_a^\infty g(t)\xi(t)dt$ as a sample function integral and as a mean square integral.

After a brief review of the existence theory of linear differential operators in Section 3, we define what is meant by a solution (in mean square) of a stochastic differential equation, and establish some results relating the covariance functions of the $\{\xi(t)\}$ and $\{z(t)\}$ processes where $\mathscr{L}z(t)=\xi(t)$. In order to deduce the representation of the covariance function of the $\{z(t)\}$ process in terms of the spectral distribution function of the (now assumed wide-sense stationary) process $\{\xi(t)\}$, we must consider various theorems concerning the interchange of order of integration in multiple improper Riemann–Stieltjes–Riemann integrals. This matter is disposed of in Section 4. In Section 5 we establish various relationships between the covariance functions, spectral distribution functions, and spectral densities of the $\{\xi(t)\}$ and $\{z(t)\}$ processes. Finally, in the last two sections, results analogous to those proved for linear differential operators are developed for linear difference operators.

1. Stochastic Integrals

Let $\{x(t) \mid t \in T\}$ be a stochastic process defined on a probability space $\mathscr{P} = (\Omega, \mathscr{F}, P)$. In Chapter II we considered various conditions under which x was continuous or differentiable with probability one, as well as conditions under which x was continuous or differentiable in mean square. We would now like to investigate under what conditions the familiar symbol

$$I = \int_T x(t)\,dt \tag{1.1}$$

is meaningful. That is, we would like to examine conditions under which I exists as a sample function integral, as well as conditions under which I exists as a mean square integral.

Our analysis is motivated by the corresponding theory in the calculus of ordinary functions in which the Riemann integral is defined as the limit of an approximating sum. We briefly recall the pertinent facts.

Let $T = [a, b]$ be a closed finite interval. Let n be a positive integer and divide the interval T into n pieces by the points of subdivision $t_0, t_1, \cdots, t_n$ where

$$a = t_0 < t_1 < \cdots < t_{n-1} < t_n = b \quad .$$

Such a subdivision is called a partition, π, of the interval T. By the norm ν of the partition π we mean the maximum of the positive numbers $t_k - t_{k-1}$, $1 \leqq k \leqq n$. In each subinterval $[t_{k-1}, t_k]$ choose an arbitrary point and call it t_k'. The partition π now becomes a marked partition, π', of T. Let x be a function defined and bounded on T. Then

$$S_{\pi'} = \sum_{k=1}^{n} x(t_k')\,(t_k - t_{k-1}) \tag{1.2}$$

is called an approximating sum for this marked partition.

Now let π' be any marked partition of norm ν and let $S_{\pi'}$ be the corresponding approximating sum. Then if there exists a number I such that given any $\varepsilon > 0$ there exists a $\delta > 0$ such that

$$|S_{\pi'} - I| \leqq \varepsilon$$

whenever $\nu < \delta$ and for all markings of the partition, then we shall say x is integrable on T and write

$$I = \int_a^b x(t)\,dt = \lim_{\nu \to 0} S_{\pi'} \quad .$$

In particular, if x is continuous on T, then I exists.

Now let x be continuous on $T = [a, b]$. Let $\{\nu_n\}$ be an arbitrary sequence of positive numbers converging to zero. Let π_n' be any marked partition of T of norm less than ν_n and let Σ_n be the corresponding approximating sum. Then the sequence $\{\Sigma_n\}$ converges to $\int_a^b x(t)dt$:

$$\lim_{n \to \infty} \Sigma_n = \int_a^b x(t)dt \quad .$$

For the remainder of this section $S_{\pi'}$ will always signify an approximating sum for an arbitrary marked partition π', while $\{\Sigma_n\}$ will refer to an arbitrary sequence of approximating sums with norms converging to zero.

Let $\{x(t) \mid t \in T\}$ be a real or complex stochastic process whose parameter set T is a closed finite interval. Then if π' is a marked partition of T, the approximating sum

$$S_{\pi'} = \sum_{k=1}^{n} x(t_k')\,(t_k - t_{k-1})$$

is a random variable. If $S_{\pi'}$ converges with probability one as the norm ν of the partition approaches zero, then the limit will be called the *sample function integral* of x and we shall write

$$\lim_{\nu \to 0} S_{\pi'} = [I]_{\text{ae}} = \left[\int_T x(t)\,dt\right]_{\text{ae}} \quad . \tag{1.3}$$

If $S_{\pi'}$ converges in mean square as the norm ν of the partition approaches zero, then the limit will be called the *mean square integral* of x and we shall write

$$\underset{\nu \to 0}{\text{l.i.m.}}\ S_{\pi'} = [I]_{\text{ms}} = \left[\int_T x(t)dt\right]_{\text{ms}} \quad . \tag{1.4}$$

If $[I]_{\text{ae}}$ exists, then an arbitrary sequence of approximating sums $\{\Sigma_n\}$ (with norms converging to zero) converges with probability one and

$$\lim_{n \to \infty} \Sigma_n = [I]_{\text{ae}} \quad . \tag{1.5}$$

Similarly, if $[I]_{\text{ms}}$ exists, then $\{\Sigma_n\}$ converges in mean square and

$$\underset{n \to \infty}{\text{l.i.m.}}\ \Sigma_n = [I]_{\text{ms}} \quad . \tag{1.6}$$

Now suppose $\lim_{n\to\infty} \Sigma_n$ and $\text{l.i.m.}_{n\to\infty}\Sigma_n$ both exist. Then by Corollary 7.1 of Chapter I, the limits are equal with probability one. That is, $[I]_{\text{ae}}$ and $[I]_{\text{ms}}$ are equivalent random variables and (cf. page 71) we may drop the "ae" and "ms" subscripts.

We shall now prove some theorems which yield sufficient conditions under which the integral of a stochastic process exists.

Theorem 1.1. Let T be a closed finite interval and let $\{x(t)\,|\,t \in T\}$ be a real or complex stochastic process whose sample functions are continuous with probability one. Then

$$\int_T x(t)dt$$

exists as a sample function integral.

Proof. Since the sample functions are continuous with probability one, $I = \int_T x(t)dt$ exists as an ordinary Riemann integral for almost all sample functions. Now let $\{\Sigma_n\}$ be an arbitrary sequence of approximating sums with norms converging to zero. We shall show that

$$\lim_{n\to\infty} \Sigma_n = I \quad ,$$

and hence $I = [I]_{\text{ae}}$ [see (1.3) and (1.5)].

Let $\varepsilon > 0$ be assigned. Then there exists an N such that, except on an ω-set of P-measure zero,

$$|\Sigma_n - I| \leqq \varepsilon$$

for all $n \geqq N$. Thus

$$\lim_{N\to\infty} P[|\Sigma_n - I| \leqq \varepsilon \quad \text{for all} \quad n \geqq N] = 1$$

and [see (4.6) of Chapter I] the sequence $\{\Sigma_n\}$ converges with probability one to I.

In terms of conditions on the covariance function of the stochastic process we have

Theorem 1.2. Let T be a closed finite interval and let $\{x(t)\,|\,t \in T\}$ be a real stochastic process with mean zero and covariance function $R(t, s) = \mathscr{E}x(t)x(s)$. For some $\alpha > 0$ let

$$R(t + h,\ t + h) - R(t + h, t) - R(t, t + h) + R(t, t) = 0(|h|^{1+\alpha})$$

as h approaches zero. Then there exists an equivalent stochastic process

$\{x^*(t)\,|\,t \in T\}$ with the property that

$$\int_T x^*(t)dt$$

exists as a sample function integral.

Proof. By the Tschebyscheff inequality

$$P[\,|x(t+h) - x(t)\,| \geqq g(h)] \leqq \frac{1}{g^2(h)}\, \mathscr{E}[x(t+h) - x(t)]^2$$

and

$$\mathscr{E}[x(t+h) - x(t)]^2 = R(t+h, t+h) - R(t+h, t) - R(t, t+h) + R(t,t) \quad .$$

Thus there exists a constant K independent of h such that

$$\mathscr{E}[x(t+h) - x(t)]^2 < K|h|^{1+\alpha} \quad .$$

The proof now proceeds as in Theorem 6.2 of Chapter II and establishes the existence of an equivalent process $\{x^*(t)\}$ whose sample functions are continuous with probability one. Theorem 1.1 completes the proof.

For mean square integrability we need only the continuity of the covariance function.

Theorem 1.3. Let $T = [a, b]$ be a closed finite interval and let $\{\xi(t)\,|\,t \in T\}$ be a real or complex stochastic process with mean zero and covariance function $R(t, s) = \mathscr{E}\xi(t)\bar{\xi}(s)$. Let R be continuous on $T \times T$. Then $\xi(t)$ is mean square integrable on T.

Proof. Let $\{\Sigma_n\}$ and $\{\Sigma_n'\}$ be two sequences of approximating sums with norms converging to zero. Then

$$\mathscr{E}\,\Sigma_n \bar{\Sigma}_m' = \sum_{k=1}^{n} \sum_{j=1}^{m} R(t_k', s_j')\,(t_k - t_{k-1})\,(s_j - s_{j-1}) \quad . \tag{1.7}$$

As n and m increase without limit, the right-hand side of the equation above approaches

$$\int_a^b \int_a^b R(t, s)\; dtds$$

by the continuity of R on $T \times T$. By Loève's criterion (Theorem 6.2 of Chapter I) we conclude that the sequence $\{\Sigma_n\}$ converges in mean square to [see (1.4) and (1.6)]

$$[I]_{\mathrm{ms}} = \left[\int_a^b \xi(t)dt\right]_{\mathrm{ms}} .$$

From Corollary 6.1 of Chapter I we have the useful consequence:

Corollary 1.1. Let the covariance function R of $\{\xi(t) \mid t \in T\}$ be continuous on $T \times T$. Then

$$\mathscr{E}[I]_{\mathrm{ms}} = 0$$

and

$$\mathscr{E}|[I]_{\mathrm{ms}}|^2 = \int_a^b \int_a^b R(t, s)\, dtds .$$

Combining Theorems 1.2 and 1.3 leads to:

Theorem 1.4. Let T be a closed finite interval and let $\{x(t) \mid t \in T\}$ be a real stochastic process with mean zero and covariance function R. For some $\alpha > 0$ let

$$R(t + h, t + h) - R(t + h, t) - R(t, t + h) + R(t, t) = 0(|h|^{1+\alpha}) \qquad (1.8)$$

as h approaches zero. Then the integral of x exists both as a sample function integral and as a mean square integral, and the two integrals are equal with probability one.

Proof. By Theorem 1.2 there exists an equivalent process $\{x^*(t) \mid t \in T\}$ whose sample functions are continuous with probability one. Thus

$$\left[\int_T x^*(t)dt\right]_{\mathrm{ae}}$$

exists as an ordinary Riemann integral for almost all sample functions x^*, and

$$\left[\int_T x(t)dt\right]_{\mathrm{ae}} = \left[\int_T x^*(t)dt\right]_{\mathrm{ae}}$$

with probability one. Thus if $\{\Sigma_n\}$ is an arbitrary sequence of approximating sums with norms converging to zero, then by Theorem 1.1

$$\lim_{n\to\infty} \Sigma_n = \left[\int_T x(t)dt\right]_{\mathrm{ae}} .$$

The hypothesis (1.8) on the covariance function implies its continuity. (See Theorems 7.1 and 7.2 of Chapter II.) Therefore by Theorem 1.3

$$\underset{n\to\infty}{\text{l.i.m.}}\ \Sigma_n = \left[\int_T x(t)dt\right]_{\text{ms}} \quad .$$

Thus by Corollary 7.1 of Chapter I, the limits above are equal with probability one.

If the process $\{x(t) \mid t \in T\}$ is wide-sense stationary, then the condition (1.8) in Theorem 1.4 may be replaced by

$$R(h) = R(0) - 0(|h|^{1+\alpha}) \quad , \qquad \alpha > 0 \quad ,$$

as h approaches zero (see Theorem 6.2 of Chapter II).

Another case of interest occurs when the stochastic process is a complex stationary Gaussian process. For this case we have from Corollary 6.1 of Chapter II (see also Corollary 7.2 of Chapter II),

Theorem 1.5. Let $\{\xi(t) \mid -\infty < t < \infty\}$ be a complex stationary Gaussian process with mean zero and covariance function R. For some $\alpha > 0$ let

$$R(h) + R(-h) = 2R(0) - 0(|h|^{\alpha})$$

as h approaches zero. Then if T is any finite interval,

$$\int_T \xi(t)dt$$

exists both as a sample function integral and as a mean square integral, and the two integrals are equal with probability one.

If $R'(0)$ exists, then certainly

$$R(h) + R(-h) = 2R(0) - 0(|h|^{\alpha})$$

as h approaches zero for any positive α less than or equal to one. Thus we have the further weaker but more convenient criterion,

Corollary 1.2. Let $\{\xi(t) \mid -\infty < t < \infty\}$ be a complex stationary Gaussian process with mean zero and covariance function R. Let $R'(0)$ exist. Then if T is any finite interval,

$$\int_T \xi(t)dt$$

exists both as a sample function integral and as a mean square integral, and the two integrals are equal with probability one.

And by Theorem 5.2 of Chapter II we have

Corollary 1.3. Let $\{\xi(t) \mid -\infty < t < \infty\}$ be a mean zero complex stationary Gaussian process with continuous covariance function and spectral distribution function S. Let

$$\int_{-\infty}^{\infty} |f| dS(f)$$

converge. Then if T is any finite interval,

$$\int_T \xi(t) dt$$

exists both as a sample function integral and as a mean square integral, and the two integrals are equal with probability one.

In Theorem 9.2 of Chapter II we showed, under suitable hypotheses, that if $\{\xi\}$ were a mean zero complex stationary Gaussian process, then the derivative process $\{\xi'\}$ was normally distributed. We shall now demonstrate that the *integral* of a Gaussian process is also normally distributed. Precisely:

Theorem 1.6. Let $T = [a, b]$ be a closed finite interval and let $\{\xi(t) \mid t \in T\}$ be a complex Gaussian process with mean zero and covariance function $R(t, s) = \mathscr{E}\xi(t)\bar{\xi}(s)$. Let R be continuous on $T \times T$. Then

$$\int_a^b \xi(t) dt$$

is normally distributed

$$\left(0, \int_a^b \int_a^b R(t, s)\, dtds\right) .$$

Proof. Let $\{\nu_n\}$ be an arbitrary sequence of positive numbers converging to zero. Let

$$\Sigma_n = \sum_{k=1}^{n} \xi(t_k')\,(t_k - t_{k-1})$$

be an approximating sum of norm less than ν_n. Then [see (1.10) of Chapter III] Σ_n is normal $(0, \sigma_n^2)$ where

$$\sigma_n^2 = \mathscr{E}|\Sigma_n|^2 = \sum_{k=1}^{n} \sum_{j=1}^{n} R(t_k', t_j')\,(t_k - t_{k-1})(t_j - t_{j-1}) .$$

By Theorem 1.3 the sequence $\{\Sigma_n\}$ converges in mean square to $\int_a^b \xi(t)dt$, and, *a fortiori,* in probability.

Now let $S_n = \text{Re}\,\Sigma_n$ and let F_n be the distribution function of S_n. Then(see page 86),

$$F_n(x) = P[S_n \leqq x] = \frac{1}{(\frac{1}{2}\sigma_n^2)^{1/2}\sqrt{2\pi}} \int_{-\infty}^{x} \exp\left[\frac{-\frac{1}{2}u^2}{(\frac{1}{2}\sigma_n^2)}\right] du$$

and

$$\lim_{n\to\infty} F_n(x) = F(x) = \frac{1}{(\frac{1}{2}\sigma^2)^{1/2}\sqrt{2\pi}} \int_{-\infty}^{x} \exp\left[\frac{-\frac{1}{2}u^2}{(\frac{1}{2}\sigma^2)}\right] du$$

where

$$\sigma^2 = \lim_{n\to\infty} \sigma_n^2 = \int_a^b \int_a^b R(t, s)\, dtds$$

by the continuity of R. We treat the imaginary part of Σ_n similarly. Thus by Theorem 5.2 of Chapter I, $\int_a^b \xi(t)dt$ is normal $(0, \sigma^2)$.

If we examine the proof of Theorem 1.6 we see the truth of the following more general proposition: Let $\{\xi_n\}$ be a sequence of complex random variables defined on a probability space $\mathscr{P}$. Let ξ_n be normal (μ_n, σ_n^2) for all n. Then if $\{\xi_n\}$ converges in mean square to ξ, the random variable ξ is normal (μ, σ^2) where $\mu = \lim_{n\to\infty} \mu_n$ and $\sigma^2 = \lim_{n\to\infty} \sigma_n^2$. (See Corollary 6.1 of Chapter I.)

2. Generalizations and Infinite Stochastic Integrals

Let $\{\xi(t) \mid t \in T\}$ be a stochastic process, and let g be a deterministic function with domain T. Then we frequently have occasion to consider the existence of

$$\int_T g(t)\xi(t)dt \quad .$$

Furthermore, if $T = [a, b]$ is a closed finite interval and $K(t, s)$ is defined on $T \times T$, then the integral

$$\int_a^t K(t, s)\, \xi(s)ds \quad , \qquad t \in T \quad ,$$

(when it exists) is also of interest.

Clearly if g is continuous on T, then with the same conditions on the stochastic process, the theorems of Section 1 are valid with the stochastic process, x or ξ, replaced by gx or $g\xi$, respectively. For example, Theorem 1.3 and Corollary 1.1 become:

Theorem 2.1. Let $T = [a, b]$ be a closed finite interval and let $\{\xi(t) \mid t \in T\}$

be a complex stochastic process with mean zero and covariance function $R(t, s) = \mathscr{E}\xi(t)\bar{\xi}(s)$. Let R be continuous on $T \times T$, and let g be a complex-valued function of the real variable t which is continuous on T. Then

$$\eta = \int_a^b g(t)\xi(t)dt \tag{2.1}$$

exists as a mean square integral and

$$\mathscr{E}\eta = 0 \quad , \tag{2.2}$$

$$\mathscr{E}|\eta|^2 = \int_a^b \int_a^b g(t)\bar{g}(s)R(t, s)\, dtds \quad , \tag{2.3}$$

and Theorem 1.5 may be worded:

Theorem 2.2. Let $\{\xi(t) | -\infty < t < \infty\}$ be a complex stationary Gaussian process with mean zero and covariance function R. For some $\alpha > 0$ let

$$2R(0) - R(h) - R(-h) = 0(|h|^\alpha)$$

as h approaches zero. Then if T is any finite interval and g is any complex-valued function of the real variable t continuous on T, then

$$\int_T g(t)\xi(t)dt$$

exists both as a sample function integral and as a mean square integral, and the two integrals are equal with probability one.

Similarly, if K is continuous on $T \times T$, then for any fixed $t_0 \in T = [a, b]$ we may consider the approximating sum

$$S_{\pi'} = \sum_{k=1}^{n} K(t_0, s_k')\xi(s_k')(s_k - s_{k-1})$$

and conclude that

$$\zeta(t_0) = \int_a^{t_0} K(t_0, s)\xi(s)\, ds \tag{2.4}$$

exists as a mean square integral for all $t_0 \in T$. By (6.9) of Chapter I,

$$\mathscr{E}\zeta(t) = \int_a^t K(t, s)\mathscr{E}\xi(s)ds = 0$$

and by Theorem 6.3 of Chapter I,

$$r(t, s) = \mathrm{Cov}(\zeta(t), \bar{\zeta}(s)) = \mathscr{E}\zeta(t)\bar{\zeta}(s)$$

$$= \int_a^t du \int_a^s dv\ K(t, u)\, \bar{K}(s, v) R(u, v) \quad . \tag{2.5}$$

Suppose now $\partial r/\partial s$ and $\partial^2 r/\partial t \partial s$ are continuous on $T \times T$. Then by Theorem 8.7 of Chapter II, this will imply that $\zeta(t)$ has a mean square derivative on T. From the form of (2.5) we see that if appropriate differentiability conditions are imposed on K, then this will be true. Precisely, we state and prove

Theorem 2.3. Let $\{\xi(t) \mid t \in T\}$ be a mean zero complex stochastic process with covariance function R continuous on $T \times T$. Let $T = [a, b]$ be a closed finite interval, and let $K(t, s)$ be defined and continuous on $T \times T$. Then if $\partial K/\partial t$ exists and is continuous on $T \times T$,

$$\zeta(t) = \int_a^t K(t, s)\, \xi(s) ds \quad , \qquad t \in T \quad ,$$

has a mean square derivative

$$\zeta'(t) = K(t, t)\xi(t) + \int_a^t \frac{\partial}{\partial t} K(t, s)\, \xi(s) ds$$

on T.

Proof. From (2.5) by the continuity of K and $\partial K/\partial s$,

$$\frac{\partial}{\partial s} r(t, s) = \int_a^t du \int_a^s dv\ K(t, u) \frac{\partial}{\partial s} \bar{K}(s, v)\ R(u, v)$$
$$+ \bar{K}(s, s) \int_a^t K(t, u) R(u, s) du$$

and

$$\frac{\partial^2}{\partial t\, \partial s} r(t, s) = \int_a^t du \int_a^s dv \frac{\partial}{\partial t} K(t, u) \frac{\partial}{\partial s} \bar{K}(s, v)\ R(u, v)$$
$$+ K(t, t) \int_a^s \frac{\partial}{\partial s} \bar{K}(s, v)\ R(t, v) dv$$
$$+ \bar{K}(s, s) \int_a^t \frac{\partial}{\partial t} K(t, u)\ R(u, s) du$$
$$+ \bar{K}(s, s)\ K(t, t) R(t, s) \quad .$$

The conditions on R and K imply the continuity of $\partial r/\partial s$ and $\partial^2 r/\partial t \partial s$. Theorem 8.7 of Chapter II completes the proof.

Corollary 2.1. If $K(x, x) = 0$ for all x in T, then

$$\operatorname{Cov}(\zeta'(t), \bar{\zeta}'(s)) = \int_a^t du \int_a^s dv \frac{\partial}{\partial t} K(t, u) \frac{\partial}{\partial s} \bar{K}(s, v)\ R(u, v) \quad .$$

Before turning to infinite integrals, let us briefly consider the problem of expanding a random process in a series of orthogonal random variables. Towards this end let $T = [a, b]$ be a closed finite interval and let $\{\xi(t) \mid t \in T\}$ be a mean zero complex stochastic process continuous in mean square on T. Then by Theorems 7.1 and 7.2 of Chapter II, the covariance function $R(t, s) = \mathscr{E}\xi(t)\bar{\xi}(s)$ of the $\{\xi(t) \mid t \in T\}$ process is continuous on $T \times T$. Now *Mercer's theorem* states that

$$R(t, s) = \sum_{n=1}^{\infty} \lambda_n \phi_n(t) \bar{\phi}_n(s)$$

where the series converges uniformly and absolutely on $T \times T$, and where the $\{\phi_n\}$ are continuous orthonormal eigenfunctions of R corresponding to the proper values λ_n:

$$\int_a^b R(t, s) \phi_n(s) ds = \lambda_n \phi_n(t) \quad .$$

By Theorem 2.1,

$$z_n = \int_a^b \bar{\phi}_n(t) \xi(t) dt \quad , \qquad n = 1, 2, \cdots \quad ,$$

exists as a mean square integral, and [see (2.5)]

$$\begin{aligned} \mathscr{E} z_n \bar{z}_m &= \int_a^b \int_a^b \bar{\phi}_n(t) \phi_m(s) R(t, s) dt ds \\ &= \lambda_m \int_a^b \bar{\phi}_n(t) \phi_m(t) dt = \lambda_m \delta_{nm} \quad . \end{aligned}$$

Also,

$$\begin{aligned} \mathscr{E}\xi(t) \bar{z}_n &= \mathscr{E} \int_a^b \phi_n(s) \bar{\xi}(s) \xi(t) ds \\ &= \int_a^b \phi_n(s) R(t, s) ds = \lambda_n \phi_n(t) \quad . \end{aligned}$$

Since

$$R(t, t) = \sum_{n=1}^{\infty} \lambda_n |\phi_n(t)|^2$$

converges by Mercer's theorem, Corollary 6.2 of Chapter I implies that the stochastic series

$$\sum_{n=1}^{\infty} \phi_n(t) z_n$$

converges in mean square.

Now it follows that

$$\xi(t) = \underset{N\to\infty}{\text{l.i.m.}} \sum_{n=1}^{N} \phi_n(t) z_n$$

since

$$\lim_{N\to\infty} \mathscr{E} \left| \sum_{n=1}^{N} \phi_n(t) z_n - \xi(t) \right|^2 = R(t, t) - \lim_{N\to\infty} \sum_{n=1}^{N} \lambda_n |\phi_n(t)|^2$$
$$= \lim_{N\to\infty} \sum_{n=N+1}^{\infty} \lambda_n |\phi_n(t)|^2 = 0$$

uniformly in t. Thus $\xi(t)$ converges uniformly in mean square to $\sum_{n=1}^{\infty} \phi_n(t) z_n$,

$$\xi(t) = \sum_{n=1}^{\infty} \phi_n(t) z_n \quad . \tag{2.6}$$

Equation (2.6) is known as the *Karhunen–Loève expansion.*

Conversely, let $\{w_n\}$ be a family of random variables with the property that

$$\mathscr{E} w_n \bar{w}_m = \mu_n \delta_{nm} \quad .$$

Let $\{\psi_n\}$ be an orthonormal sequence of continuous functions with common domain $T = [a, b]$ such that the series

$$\sum_{n=1}^{\infty} \mu_n \psi_n(t) \bar{\psi}_n(s)$$

converges uniformly on $T \times T$. Let $\{\xi(t) \mid t \in T\}$ be a mean zero complex stochastic process with covariance function R continuous on $T \times T$. Then if

$$\xi(t) = \sum_{n=1}^{\infty} \psi_n(t) w_n \quad ,$$

the $\{\psi_n\}$ are eigenfunctions of R with corresponding proper values μ_n. For, by Theorem 6.3 of Chapter I,

$$R(t, s) = \mathscr{E}\xi(t)\bar{\xi}(s) = \mathscr{E}\left[\sum_{n=1}^{\infty} \psi_n(t) w_n \right]\left[\sum_{m=1}^{\infty} \bar{\psi}_m(s) \bar{w}_m \right]$$
$$= \sum_{n=1}^{\infty} \mu_n \psi_n(t) \bar{\psi}_n(s) \quad ,$$

and by the uniform convergence of this series,

$$\int_a^b R(t, s)\psi_m(s)ds = \sum_{n=1}^{\infty} \mu_n \psi_n(t) \int_a^b \bar{\psi}_n(s)\psi_m(s)ds$$
$$= \sum_{n=1}^{\infty} \mu_n \psi_n(t)\delta_{nm} = \mu_m \psi_m(t) \quad .$$

We now turn our attention to infinite integrals. Suppose $[a, \infty) \subset T$, and R, the covariance function of the $\{\xi(t) | t \in T\}$ process, is continuous on $[a, \infty) \times [a, \infty)$. Suppose further that g is continuous on $[a, \infty)$. Then

$$\int_a^b \int_a^{b'} g(t)\bar{g}(s)R(t, s)dtds$$

exists for all $b, b' > a$. If

$$\lim_{b,b' \to \infty} \int_a^b \int_a^{b'} g(t)\bar{g}(s)R(t, s)dtds$$

exists then we assert that

$$\int_a^{\infty} g(t)\xi(t)dt$$

exists as a mean square integral. Formally:

Theorem 2.4. Let $\{\xi(t) | -\infty < t < \infty\}$ be a complex stochastic process with mean zero and covariance function $R(t, s) = \mathscr{E}\xi(t)\bar{\xi}(s)$. Let R be continuous on $[a, \infty) \times [a, \infty)$ and let g be a complex-valued function of the real variable t continuous on $[a, \infty)$. Let

$$\lim_{b,b' \to \infty} \int_a^b \int_a^{b'} g(t)\bar{g}(s)R(t, s)dtds \tag{2.7}$$

exist. Then

$$z = \int_a^{\infty} g(t)\xi(t)dt$$

exists as a mean square integral and

$$\mathscr{E}z = 0 \quad ,$$
$$\mathscr{E}|z|^2 = \int_a^{\infty} \int_a^{\infty} g(t)\bar{g}(s)R(t, s)dtds \quad .$$

Proof. Let $\{t_n\}$ be an arbitrary strictly increasing sequence of real numbers greater than a and tending to $+\infty$. Let

$$z_n = \int_a^{t_n} g(t)\xi(t)dt \quad .$$

Then

$$\mathscr{E}|z_n - z_m|^2 = \mathscr{E}\left|\int_{t_m}^{t_n} g(t)\xi(t)dt\right|^2$$
$$= \int_{t_m}^{t_n}\int_{t_m}^{t_n} g(t)\bar{g}(s)R(t, s)dtds \quad , \qquad n > m \quad ,$$

by Theorem 2.1. We easily see that

$$\lim_{n,m\to\infty} \mathscr{E}|z_n - z_m|^2 = 0 \quad ,$$

and by the Cauchy convergence theorem for real numbers, z exists. Corollary 6.1 of Chapter I completes the proof.

Corollary 2.2. Let $\{\xi(t)| -\infty < t < \infty\}$ be a complex stochastic process with mean zero and covariance function R continuous on $[a, \infty) \times [a, \infty)$. Let g be a complex-valued function of the real variable t continuous on $[a, \infty)$. Let

$$\int_a^\infty |g(t)|(R(t, t))^{1/2}dt < \infty \quad . \tag{2.8}$$

Then

$$z = \int_a^\infty g(t)\xi(t)dt$$

exists as a mean square integral.

Proof. For all $b, b' > a$,

$$\left|\int_a^b \int_a^{b'} g(t)\bar{g}(s)R(t, s)dtds\right| \leq \int_a^b \int_a^{b'} |g(t)|\,|g(s)|[R(t, t)R(s, s)]^{1/2}\, dtds$$

since

$$|R(t, s)|^2 \leq R(t, t)R(s, s) \quad .$$

Thus

$$\left|\int_a^b \int_a^{b'} g(t)\bar{g}(s)R(t, s)dtds\right| \leq \left[\int_a^b |g(t)|(R(t, t))^{1/2}dt\right]\left[\int_a^{b'} |g(s)|(R(s, s))^{1/2}ds\right]$$

and by (2.8)

$$\lim_{b,b'\to\infty} \int_a^b \int_a^{b'} g(t)\bar{g}(s)R(t, s)dtds$$

exists. We may now invoke Theorem 2.4.

The continuity of the sample functions and the hypotheses of Corollary 2.2 are sufficient to guarantee the existence of the infinite integral as a sample function integral. That is, we shall prove that if $\{\tau_n\}$ is an arbitrary monotone sequence of real numbers diverging to $+\infty$, then

$$\lim_{n\to\infty} \int_a^{\tau_n} g(t)\xi(t)dt$$

exists.

Theorem 2.5. Let $\{\xi(t) \mid -\infty < t < \infty\}$ be a complex stochastic process with mean zero and covariance function $R(t, s) = \mathscr{E}\xi(t)\bar{\xi}(s)$. Let the sample functions be continuous with probability one, let R be continuous on $[a, \infty) \times [a, \infty)$, and let g be a complex-valued function of the real variable t continuous on $[a, \infty)$. Let

$$\int_a^{\infty} |g(t)|(R(t, t))^{1/2} dt < \infty \quad . \tag{2.9}$$

Then

$$\int_a^{\infty} g(t)\xi(t)dt$$

exists as a sample function integral.

Proof. Since

$$\int_a^{\infty} |g(t)|(R(t, t))^{1/2} dt$$

converges, there exists a $t_1 > a$ such that

$$\int_{t_1}^{\infty} |g(t)|(R(t, t))^{1/2} dt < 1$$

and a $t_n > \max(n-1, t_{n-1})$, $n = 2, 3, 4, \cdots$, such that

$$\int_{t_n}^{\infty} |g(t)|(R(t, t))^{1/2} dt < \frac{1}{n^2} \quad . \tag{2.10}$$

Let

$$z_n = \int_a^{t_n} g(t)\xi(t)dt \quad .$$

Then

$$|z_{n+1} - z_n| = \left| \int_{t_n}^{t_{n+1}} g(t)\xi(t)dt \right|$$

and

$$\mathscr{E}|z_{n+1} - z_n|^2 \leqq \int_{t_n}^{t_{n+1}} \int_{t_n}^{t_{n+1}} |g(t)||g(s)||R(t, s)|dtds$$

by Theorem 2.1. Hence

$$\mathscr{E}|z_{n+1} - z_n|^2 \leqq \left[\int_{t_n}^{\infty} |g(t)|(R(t, t))^{1/2}dt\right]^2 < \frac{1}{n^4}$$

by (2.10). From the Tschebyscheff inequality

$$P\left[|z_{n+1} - z_n| > \frac{1}{n}\right] < \frac{1}{n^2} \quad ,$$

and by Lemma 4.1 of Chapter I, the sequence $\{z_n\}$ converges with probability one to a P-measurable function z,

$$\lim_{n\to\infty} z_n = \lim_{n\to\infty} \int_a^{t_n} g(t)\xi(t)dt = \int_a^{\infty} g(t)\xi(t)dt = z \quad .$$

Now let $\{\tau_n\}$ be an arbitrary monotone increasing sequence of real numbers greater than a diverging to $+\infty$. We shall show that the sequence $\{y_n\}$ where

$$y_n = \int_a^{\tau_n} g(t)\xi(t)dt$$

also converges with probability one to z. This will then establish the fact that $\int_a^\infty g(t)\xi(t)dt$ exists as a sample function integral.

First we note that if $\beta > b > a$, then

$$\mathscr{E}\left|\int_b^\beta g(t)\xi(t)dt\right| \leqq \mathscr{E}\int_b^\beta |g(t)||\xi(t)|dt \leqq \int_b^\beta |g(t)|(R(t, t))^{1/2}dt \qquad (2.11)$$

since

$$\mathscr{E}^2|\xi(t)| \leqq R(t, t) \quad .$$

Now suppose

$$t_n \leqq \tau_{k_n+1} \leqq \tau_{k_n+2} \leqq \cdots \leqq \tau_{k_{n+1}} < t_{n+1}$$

and $t_{n+1} \leqq \tau_{k_{n+1}+1}$ for all n. Let

$$Y_n = \max_{k_n < j \leqq k_{n+1}} |y_j - z| \quad .$$

Then

$$\mathscr{E}Y_n < \frac{1}{n^2} \quad .$$

By the Markov inequality

$$P[Y_n > \varepsilon] \leqq \frac{\mathscr{E}Y_n}{\varepsilon}$$

for every $\varepsilon > 0$. Thus by virtue of (4.13) of Chapter I, the sequence $\{Y_n\}$ converges with probability one to zero, and hence $\{y_n\}$ converges with probability one to z.

Since $\{y_n\}$ also converges in mean square to $\int_a^\infty g(t)\xi(t)dt$ by Corollary 2.2 and Theorem 2.4, we conclude from Corollary 7.1 of Chapter I that:

Corollary 2.3. Under the hypotheses of Theorem 2.5,

$$\int_a^\infty g(t)\xi(t)dt$$

exists both as a sample function integral and as a mean square integral, and the two integrals are equal with probability one.

The results of this section may readily be extended to include integrals of the form $\int_{-\infty}^a g(t)\xi(t)dt$ and $\int_{-\infty}^\infty g(t)\xi(t)dt$. In particular we shall find the following theorem useful in future work.

Theorem 2.6. Let $\{\xi(t) \mid -\infty < t < \infty\}$ be a mean zero complex stationary Gaussian process with continuous covariance function R. Let $R'(0)$ exist. Let g be a complex-valued function of the real variable t continuous on $(-\infty,\infty)$. Let g be absolutely integrable on $(-\infty,\infty)$. Then

$$\int_{-\infty}^\infty g(t)\xi(t)dt$$

exists both as a sample function integral and as a mean square integral, and the two integrals are equal with probability one.

Proof. The existence of $R'(0)$ implies that

$$2R(0) - R(h) - R(-h) = 0(|h|^\alpha)$$

as h approaches zero for some $\alpha > 0$. Thus by Corollary 6.1 of Chapter II there exists a complex process equivalent to $\{\xi(t)\}$ which possesses continuous sample functions with probability one. Since $R(t, t) = R(0)$ for a wide-sense stationary process,

$$\int_{-\infty}^{\infty} |g(t)|(R(t,t))^{1/2}\,dt = (R(0))^{1/2}\int_{-\infty}^{\infty}|g(t)|\,dt < \infty \quad .$$

Thus by Theorem 2.5,

$$\int_{-\infty}^{\infty} g(t)\xi(t)dt$$

exists as a sample function integral. An appeal to Corollary 2.3 completes the proof.

In terms of the spectral distribution function (cf. Theorem 5.2 of Chapter II),

Theorem 2.7. Let $\{\xi(t)\,|\,-\infty < t < \infty\}$ be a mean zero complex stationary Gaussian process with continuous covariance function R and spectral distribution function S. Let

$$\int_{-\infty}^{\infty} |f|\,dS(f)$$

converge. Let g be a complex-valued function of the real variable t continuous on $(-\infty, \infty)$ and let

$$\int_{-\infty}^{\infty} |g(t)|\,dt < \infty \quad .$$

Then

$$\int_{-\infty}^{\infty} g(t)\xi(t)dt$$

exists both as a sample function integral and as a mean square integral, and the two integrals are equal with probability one.

Finally, we prove an ergodic property of stochastic processes. Let $\{\xi(t)\,|\,-\infty < t < \infty\}$ be a complex stochastic process with mean μ independent of t, and continuous covariance function R. Then for any $T > 0$,

$$\mathscr{E}\left|\frac{1}{T}\int_{-\frac{1}{2}T}^{\frac{1}{2}T}[\xi(t)-\mu]dt\right|^2 = \frac{1}{T^2}\int_{-\frac{1}{2}T}^{\frac{1}{2}T}\int_{-\frac{1}{2}T}^{\frac{1}{2}T} R(t,s)dtds \tag{2.12}$$

by Theorem 2.1 (or Corollary 1.1). Now if

$$\lim_{T\to\infty}\frac{1}{T^2}\int_{-\frac{1}{2}T}^{\frac{1}{2}T}\int_{-\frac{1}{2}T}^{\frac{1}{2}T} R(t,s)dtds = 0 \quad , \tag{2.13}$$

then (2.12) implies that

$$\frac{1}{T}\int_{-\frac{1}{2}T}^{\frac{1}{2}T} \xi(t)dt$$

converges in mean square to μ. But by hypothesis

$$\mathscr{E}\xi(t) = \mu \quad .$$

Thus if (2.13) holds,

$$\underset{T\to\infty}{\text{l.i.m.}} \frac{1}{T}\int_{-\frac{1}{2}T}^{\frac{1}{2}T} \xi(t)dt = \mathscr{E}\xi(t) \quad .$$

That is, "time averages" and "ensemble averages" are equal.

3. Linear Differential Equations

Let

$$\mathscr{L} = p_0(t)\frac{d^n}{dt^n} + p_1(t)\frac{d^{n-1}}{dt^{n-1}} + \cdots + p_n(t)\cdot \tag{3.1}$$

be an nth-order linear differential operator whose coefficients p_j, $0 \leqq j \leqq n$, are continuous complex-valued functions of the real variable t on the closed interval $T = [a, b]$, and let $p_0(t) \neq 0$ on T. Then if $\{\xi(t) | t \in T\}$ is a random process, the equation

$$\mathscr{L} z(t) = \xi(t) \tag{3.2}$$

defines a new random process $\{z(t) | t \in T\}$. Equation (3.2) is analogous to the definition of autoregressive processes in the theory of time series (see Sections 6 and 7.)

Before considering the solution of (3.2) where $\xi(t)$ is a random variable, let us examine a few relevant facts concerning the solution of nonhomogeneous linear differential equations when the forcing function is an ordinary function. First we recall that the homogeneous equation

$$\mathscr{L}\phi(t) = 0$$

has n linearly independent solutions, say $\phi_1, \cdots, \phi_n$, which are continuous and have continuous derivatives of order n on T. The one-sided Green's function H of $\mathscr{L}$ is defined as

$$H(t, s) = \frac{(-1)^{n-1}}{p_0(s)W(s)} \begin{vmatrix} \phi_1(t) & \phi_2(t) & \cdots & \phi_n(t) \\ \phi_1(s) & \phi_2(s) & \cdots & \phi_n(s) \\ \phi_1'(s) & \phi_2'(s) & \cdots & \phi_n'(s) \\ \vdots & \vdots & & \vdots \\ \phi_1^{(n-2)}(s) & \phi_2^{(n-2)}(s) & \cdots & \phi_n^{(n-2)}(s) \end{vmatrix}$$

where

$$W(s) = \begin{vmatrix} \phi_1(s) & \phi_2(s) & \cdots & \phi_n(s) \\ \phi_1'(s) & \phi_2'(s) & \cdots & \phi_n'(s) \\ \vdots & \vdots & & \vdots \\ \phi_1{}^{(n-1)}(s) & \phi_2{}^{(n-1)}(s) & \cdots & \phi_n{}^{(n-1)}(s) \end{vmatrix}$$

is the Wronskian. Since

$$W(s) = W_0 \exp\left[-\int_a^s \frac{p_1(t)}{p_0(t)}\,dt\right]$$

where W_0 is a nonzero constant, we see that W is never zero on T.

Some elementary properties of the Green's function are

(i) $H(t, s)$ is unique.

(ii) $H(t, s)$ and $\frac{\partial^k}{\partial t^k} H(t, s)$, $1 \leqq k \leqq n$, are continuous on $T \times T$. (3.3)

(iii) $\left.\frac{\partial^k}{\partial t^k} H(t, s)\right|_{t=s} = 0 \quad \text{for} \quad 0 \leqq k \leqq n-2.$ (3.4)

(iv) $\left.\frac{\partial^{n-1}}{\partial t^{n-1}} H(t, s)\right|_{t=s} = \frac{1}{p_0(s)}$.

For further details see [CoL], [Mil 8].

One may now readily verify that if ψ is a function continuous on $[a, b]$, then

$$\phi(t) = \int_a^t H(t, s)\,\psi(s)ds$$

is the unique solution of

$$\mathscr{L}\phi(t) = \psi(t)$$

which assumes the initial values $\phi^{(k)}(a) = 0, 0 \leqq k \leqq n-1$.

Suppose now $\{\xi(t) \mid -\infty < t < \infty\}$ is a mean zero complex stochastic process with continuous covariance function R. Then

$$z_a(t) = \int_a^t H(t, s)\xi(s)ds \quad , \qquad t \in T \quad , \tag{3.5}$$

exists as a mean square integral [see (2.4)] and its covariance function [see (2.5)] is

$$r_a(t, s) = \int_a^t du \int_a^s dv \;\; H(t, u)\bar{H}(s, v)R(u, v) \quad .$$

Furthermore, since $\partial H/\partial t$ is continuous by (3.3) and $H(t, t) = 0$ by (3.4), Theorem 2.3 implies that

$$z_a'(t) = \int_a^t \frac{\partial}{\partial t} H(t, s)\, \xi(s)ds$$

exists as a mean square integral and (see Corollary 2.1) its covariance function is

$$\rho(t, s) = \frac{\partial^2}{\partial t\, \partial s} r_a(t, s) = \int_a^t du \int_a^s dv \frac{\partial}{\partial t} H(t, u) \frac{\partial}{\partial s} \bar{H}(s, v)\, R(u, v) \quad .$$

But the condition (3.3) on the Green's function implies that $\partial \rho/\partial s$ and $\partial^2 \rho/\partial t \partial s$ are continuous on $T \times T$. Hence Theorem 8.7 of Chapter II implies that z_a' has a mean square derivative on T. Similarly we establish the existence (in mean square) of the higher-order derivatives of z_a. Combining these facts leads to:

Theorem 3.1. Let $\{\xi(t) \mid -\infty < t < \infty\}$ be a mean zero complex stochastic process with continuous covariance function R. Let

$$\mathscr{L} = p_0(t) \frac{d^n}{dt^n} + p_1(t) \frac{d^{n-1}}{dt^{n-1}} + \cdots + p_n(t) \cdot$$

be an nth-order linear differential operator whose coefficients p_j, $0 \leqq j \leqq n$, are continuous complex-valued functions of the real variable t on $T = [a, b]$, and $p_0(t) \neq 0$ for all $t \in T$. Then if H is the one-sided Green's function for $\mathscr{L}$,

$$z_a(t) = \int_a^t H(t, s)\xi(s)ds$$

exists as a mean square integral, and is the solution of the differential equation

$$\mathscr{L} z_a(t) = \xi(t)$$

where $z_a^{(k)}(a) = 0$, $0 \leqq k \leqq n-1$, with probability one. The $\{z_a(t) \mid t \in T\}$ process has mean zero and covariance function

$$r_a(t, s) = \int_a^t du \int_a^s dv\, H(t, u)\bar{H}(s, v)R(u, v) \quad . \tag{3.6}$$

A trivial generalization is

Corollary 3.1. If g is a complex-valued function of the real variable t continuous on $[a, b]$, then

$$\mathscr{L} z(t) = g(t)\xi(t)$$

defines a mean zero complex stochastic process with covariance function

$$c(t, s) = \int_a^t du \int_a^s dv\, H(t, u)\bar{H}(s, v)g(u)\bar{g}(v)R(u, v) \quad . \tag{3.7}$$

In order to avoid certain problems associated with causal systems we shall let a recede to negative infinity, so that in effect (3.5) and (3.6) are independent of a. Let us see what conditions we must impose on H and R for

$$z(t) = \int_{-\infty}^t H(t, s)\xi(s)ds$$

to be a valid process. Towards this end let $t'' < t' < t$ and $s'' < s' < s$. Then from (3.6)

$$\begin{aligned} &\left| \int_{t''}^{t'} du \int_{s''}^{s'} dv\, H(t', u)\bar{H}(s', v)R(u, v) \right| \\ &\leqq \int_{t''}^{t'} du \int_{s''}^{s'} dv\, |H(t', u)|\,|H(s', v)|\,|R(u, v)| \\ &\leqq \int_{t''}^{t'} |H(t', u)|(R(u, u))^{1/2} du \int_{s''}^{s'} |H(s', v)|(R(v, v))^{1/2} dv \quad . \end{aligned} \tag{3.8}$$

Hence if $|H(t, u)|(R(u, u))^{1/2}$ is integrable on $(-\infty, t]$ for all t, the integral

$$\int_{-\infty}^t du \int_{-\infty}^s dv\, H(t, u)\bar{H}(s, v)R(u, v)$$

exists, and by Theorem 2.4,

$$z(t) = \int_{-\infty}^t H(t, s)\xi(s)ds$$

exists as a mean square integral. Similarly, the existence of

$$\int_{-\infty}^t \left| \frac{\partial}{\partial t} H(t, u) \right| (R(u, u))^{1/2} du$$

implies that the derivative process

$$z'(t) = \int_{-\infty}^t \frac{\partial}{\partial t} H(t, s)\, \xi(s)ds$$

exists as a mean square integral. Thus with the obvious conditions on the higher partial derivatives of H with respect to t, we establish the existence (in mean square) of the higher-order derivatives of z. Clearly

$$\mathscr{L} z(t) = \xi(t) \quad .$$

Formally stated, we have the following theorem.

Theorem 3.2. Let $\{\xi(t) \mid -\infty < t < \infty\}$ be a mean zero complex stochastic process with continuous covariance function R. Let H be the one-sided Green's function of the nth-order linear differential operator $\mathscr{L}$ with continuous coefficients on $(-\infty, \infty)$ and with leading coefficient nonzero for all t. Then if $\partial^k H(t, u)/\partial t^k \, (R(u, u))^{1/2}$, $0 \leqq k \leqq n - 1$, is absolutely integrable on $(-\infty, t]$ for all t,

$$r(t, s) = \int_0^\infty \int_0^\infty H(t, t - x)\bar{H}(s, s - y)R(t - x, s - y)dxdy \tag{3.9}$$

is the covariance function of the mean zero complex stochastic process $\{z(t) \mid -\infty < t < \infty\}$ determined by the stochastic differential equation

$$\mathscr{L} z(t) = \xi(t) \quad .$$

Now let us make the additional assumption that $\mathscr{L}$ has *constant coefficients,* say

$$\mathscr{L}\phi(t) = p_0\phi^{(n)}(t) + p_1\phi^{(n-1)}(t) + \cdots + p_n\phi(t) \tag{3.10}$$

where the p_j, $0 \leqq j \leqq n$, are complex constants and $p_0 \neq 0$. Then the Green's function H of $\mathscr{L}$ is a function only of the difference of its arguments, $H(t, s) = H(t - s)$, and (3.9) becomes

$$r(t, s) = \int_0^\infty \int_0^\infty H(x)\bar{H}(y)R(t - x, s - y)dxdy \quad . \tag{3.11}$$

Finally we make the further assumption that $\{\xi(t) \mid -\infty < t < \infty\}$ is wide-sense stationary. Then the covariance function R of $\{\xi(t)\}$ is a function only of the difference of its arguments, $R(u, v) = R(u - v)$, and (3.11) becomes

$$r(t, s) = \int_0^\infty \int_0^\infty H(x)\bar{H}(y)R(t - s - x + y)dxdy \quad . \tag{3.12}$$

Note that $r(t, s)$ is now a function only of $t - s$. Hence the stochastic differential equation $\mathscr{L} z(t) = \xi(t)$ defines a wide-sense stationary process $\{z(t) \mid -\infty < t < \infty\}$.

From (3.12)

$$|r(t - s)| \leq R(0)\left[\int_0^\infty |H(x)|\,dx\right]^2 \quad ,$$

and the condition in Theorem 3.2 on the integrability of

$$\frac{\partial^k}{\partial t^k} H(t, u)\,(R(u, u))^{1/2}$$

reduces to the condition that $H^{(k)}$ be absolutely integrable on $[0, \infty)$. Since all solutions of the homogeneous equation $\mathscr{L}\phi(t) = 0$ are of the form $t^k e^{ct}$ when $\mathscr{L}$ has constant coefficients, we see that a necessary and sufficient condition that $|H^{(k)}(x)|$ be integrable on $[0, \infty)$ is that all the zeros of the indicial polynomial [see (3.10)]

$$Q(\lambda) = p_0\lambda^n + p_1\lambda^{n-1} + \cdots + p_n$$

have negative real parts.

Summarizing these special cases we have:

Theorem 3.3. Let $\{\xi(t) \mid -\infty < t < \infty\}$ be a wide-sense stationary mean zero complex stochastic process with continuous covariance function R. Let H be the one-sided Green's function of the linear differential operator

$$\mathscr{L} = p_0 \frac{d^n}{dt^n} + p_1 \frac{d^{n-1}}{dt^{n-1}} + \cdots + p_n \cdot$$

where the p_j, $0 \leqq j \leqq n$, are complex constants and $p_0 \neq 0$. Let all the zeros of the indicial polynomial

$$Q(\lambda) = p_0\lambda^n + p_1\lambda^{n-1} + \cdots + p_n$$

have negative real parts. Then the stochastic differential equation

$$\mathscr{L} z(t) = \xi(t)$$

defines a wide-sense stationary mean zero complex stochastic process $\{z(t) \mid -\infty < t < \infty\}$ with covariance function

$$r(h) = \int_0^\infty \int_0^\infty H(x)\bar{H}(y)R(h - x + y)dxdy \quad . \tag{3.13}$$

4. Riemann–Stieltjes Integrals

Let $\{\xi(t) \mid -\infty < t < \infty\}$ be a wide-sense stationary complex stochastic process with mean zero. Let the covariance function R of the process be continuous. Then by Bochner's theorem (Theorem 5.1 of Chapter II) the spectral distribution function S of the $\{\xi(t)\}$ process exists and

$$R(t) = \int_{-\infty}^{\infty} e^{i2\pi tf}\, dS(f) \quad .$$

Hence we may write (3.13) as

$$r(h) = \int_0^\infty \int_0^\infty H(x)\bar{H}(y)\left[\int_{-\infty}^{\infty} e^{i2\pi(h-x+y)f}\, dS(f)\right]dxdy \quad . \tag{4.1}$$

Assuming for the moment the validity of interchanging the order of integration in (4.1), there results

$$r(h) = \int_{-\infty}^{\infty} |M(\lambda)|^2 e^{i2\pi h\lambda} dS(\lambda) \tag{4.2}$$

where

$$M(\lambda) = \int_{0}^{\infty} H(x) e^{-i2\pi\lambda x} dx \quad . \tag{4.3}$$

Equation (4.2) is a useful formula in both theoretical and practical investigations. We shall therefore devote this section to a brief treatment of Riemann–Stieltjes integrals in order to establish sufficient conditions under which (4.2) holds (see [Wi]).

Let G be monotone increasing and bounded on $(-\infty, \infty)$. Then we define $G(\infty)$ and $G(-\infty)$ as

$$G(\infty) = \lim_{x \to \infty} G(x)$$

and

$$G(-\infty) = \lim_{x \to -\infty} G(x) \quad ,$$

respectively. Now let f be continuous on $[a, \infty)$ and let G be monotone increasing on $[a, \infty)$. Then if

$$\lim_{b \to \infty} \int_{a}^{b} f(x) dG(x)$$

exists, we write this limit as

$$\int_{a}^{\infty} f(x) dG(x) \tag{4.4}$$

and call it an improper Riemann–Stieltjes integral (see Section 3 of Chapter I).

Let f be continuous and bounded on $[a, \infty)$, and let G be monotone increasing and bounded on $[a, \infty)$. Then given any $\varepsilon > 0$ there exists an $A > a$ such that for all $b' > b > A$,

$$\int_{b}^{b'} dG(x) < \varepsilon \quad .$$

Thus for $b' > b > A$,

$$\left| \int_{a}^{b'} f(x) dG(x) - \int_{a}^{b} f(x) dG(x) \right| \leq M \int_{b}^{b'} dG(x) < M \varepsilon$$

where $M \geqq |f(x)|$ on $[a, \infty)$. Hence in this case we conclude by the Cauchy convergence theorem for real numbers that (4.4) exists. Further, we note that if f is continuous and absolutely integrable on $(-\infty, \infty)$, then its Fourier transform

$$F(\lambda) = \int_{-\infty}^{\infty} f(x) e^{-i2\pi\lambda x} dx$$

exists and is uniformly continuous and bounded on $(-\infty, \infty)$.

We shall now prove our basic theorem which yields sufficient conditions under which the order of integration may be interchanged in Riemann–Stieltjes–Riemann integrals with finite limits.

Theorem 4.1. Let f be continuous on the rectangle $\mathscr{R} = [a, b] \times [\alpha, \beta]$. Let G be bounded and monotone increasing on $[\alpha, \beta]$. Then

$$\int_a^b dx \int_\alpha^\beta f(x, y)\, dG(y) = \int_\alpha^\beta dG(y) \int_a^b f(x, y)\, dx \quad .$$

That is, both integrals exist as Riemann–Stieltjes integrals, and they are equal.

Proof. Since $f(x, y)$ is continuous for every $y \in [\alpha, \beta]$,

$$F(x, y) = \int_a^x f(x, y) dx \tag{4.5}$$

exists for every y in $[\alpha, \beta]$ and

$$\frac{\partial F}{\partial x} = f(x, y)$$

for all $(x, y) \in \mathscr{R}$. Furthermore, we assert that F is continuous on $\mathscr{R}$. To prove this, write

$$\begin{aligned} &F(x + h, y + k) - F(x, y) \\ &\qquad = \int_a^x [f(x, y + k) - f(x, y)]\, dx + \int_x^{x+h} f(x, y + k)\, dx \quad . \end{aligned}$$

Since f is continuous on $\mathscr{R}$, it is bounded and uniformly continuous on $\mathscr{R}$. Therefore, for h and k sufficiently small, the expression above may be made less (in absolute value) than any preassigned positive number.

The continuity of F implies the existence of

$$\phi(x) = \int_\alpha^\beta F(x, y)\, dG(y) \tag{4.6}$$

for every $x \in [a, b]$. Now since $\partial F/\partial x$ exists (and is continuous) we may use the Law of the mean to write

$$\frac{\phi(x+h)-\phi(x)}{h} = \int_\alpha^\beta \frac{\partial}{\partial x} F(x, y)\, dG(y) + \int_\alpha^\beta \left[\frac{\partial}{\partial x} F(x+\theta h, y) - \frac{\partial}{\partial x} F(x, y)\right] dG(y)$$

where $0 < \theta < 1$. Using the boundedness of G on $[\alpha, \beta]$ and the continuity of $\partial F/\partial x$, we infer the existence of ϕ' for all x in $[a, b]$. Hence we may write

$$\frac{d\phi}{dx} = \int_\alpha^\beta \frac{\partial}{\partial x} F(x, y)\, dG(y) = \int_\alpha^\beta f(x, y)\, dG(y) \quad . \tag{4.7}$$

Again appealing to the continuity of f and the boundedness of G we may show that $d\phi/dx$ is continuous on $[a, b]$.

If we integrate (4.7),

$$\phi(b) - \phi(a) = \int_a^b \frac{d\phi}{dx} dx = \int_a^b dx \int_\alpha^\beta f(x, y)\, dG(y) \quad , \tag{4.8}$$

while from (4.5)

$$F(b, y) - F(a, y) = \int_a^b f(x, y) dx \quad .$$

Thus by (4.6)

$$\phi(b) - \phi(a) = \int_\alpha^\beta dG(y) \int_a^b f(x, y)\, dx \quad . \tag{4.9}$$

Comparing (4.8) and (4.9) we see the truth of the theorem.

A generalization of the above theorem is given in the following corollary.

Corollary 4.1. Let $\Phi(x, y, \lambda)$ be continuous on the closed rectangular parallelepiped $\mathscr{B} = I \times I' \times J$. Let $G(\lambda)$ be bounded and monotone increasing on J. Then the iterated integral of $\Phi(x, y, \lambda)$ on $\mathscr{B}$ with respect to x, y, $G(\lambda)$ is independent of the order of integration.

Proof. If $\psi(x, y)$ is continuous on the closed rectangle $\mathscr{R} = I \times I'$, then we recall that

$$\int_{\mathscr{R}}\!\!\int \psi(x, y)\, d\mathscr{R} = \int_I dx \int_{I'} \psi(x, y)\, dy = \int_{I'} dy \int_I \psi(x, y)\, dx \quad .$$

Repeated applications of Theorem 4.1 and this result prove the corollary.

We now turn our attention to iterated infinite integrals. If

$$\lim_{b,b'\to\infty} \int_a^b dx \int_\alpha^{b'} f(x, y)\, dG(y) \tag{4.10}$$

exists, we shall write it as

$$L = \int_a^\infty dx \int_\alpha^\infty f(x, y)\, dG(y) \quad . \tag{4.11}$$

Similarly, if

$$\lim_{b,b'\to\infty} \int_\alpha^{b'} dG(y) \int_a^b f(x, y)\, dx \tag{4.12}$$

exists, we shall write it as

$$L' = \int_\alpha^\infty dG(y) \int_a^\infty f(x, y)\, dx \quad . \tag{4.13}$$

Under suitable conditions the existence of L or L' implies the existence of the other. More precisely:

Theorem 4.2. Let f be continuous on $[a, \infty) \times [\alpha, \infty)$ and let G be bounded and monotone increasing on $[\alpha, \infty)$. Then if either L or L' exists, the other exists and they are equal.

Proof. Suppose L exists. Let $\varepsilon > 0$ be assigned. Choose an $A > \max(a, \alpha)$ such that for all $b, b' > A$,

$$\left| \int_a^b dx \int_\alpha^{b'} f dG - L \right| < \varepsilon \quad .$$

Now

$$\int_a^b dx \int_\alpha^{b'} f dG = \int_\alpha^{b'} dG \int_a^b f dx$$

by Theorem 4.1. Therefore L' exists and equals L. Conversely, if L' exists, we see that L also exists and that the two integrals are equal.

A sufficient condition that insures the existence of L or L' for a certain class of functions is given in Theorem 4.3.

Theorem 4.3. Let f be continuous and absolutely integrable on $[a, \infty)$. Let h be continuous and bounded on $[\alpha, \infty)$. Let G be bounded and monotone increasing on $[\alpha, \infty)$. Then, if β and t are real,

$$\lim_{b,b' \to \infty} \int_a^b f(x)\,dx \int_\alpha^{b'} h(y) e^{i(\beta x+t)y} dG(y)$$

exists.

Proof. Consider the identity

$$\int_a^B f(x)\,dx \int_\alpha^{B'} h(y)e^{i(\beta x+t)y}dG(y) - \int_a^b f(x)\,dx \int_\alpha^{b'} h(y)e^{i(\beta x+t)y}dG(y)$$

$$= \int_b^B f(x)\,dx \int_\alpha^{B'} h(y)e^{i(\beta x+t)y}dG(y) + \int_a^B f(x)\,dx \int_{b'}^{B'} h(y)e^{i(\beta x+t)y}dG(y) \ .$$

Now since f is absolutely integrable on $[a, \infty)$, let

$$\int_a^\infty |f(x)|\,dx = \| f \| < \infty \quad .$$

Since G is bounded on $[\alpha, \infty)$, there exists a finite constant M such that

$$M = \int_\alpha^\infty dG(y) = G(\infty) - G(\alpha) \quad .$$

Furthermore, since h is bounded, let

$$|h(y)| < N$$

for all $y \in [\alpha, \infty)$.

Let $\varepsilon > 0$ be assigned, and let $K = \max(M, \| f \|)$. Now choose an $A >$ max (a, α) such that for all $C > A$,

$$\int_C^\infty |f(x)|\,dx < \frac{\varepsilon}{2KN}$$

and

$$\int_C^\infty dG(y) < \frac{\varepsilon}{2KN} \quad .$$

Then for all $b, b', B, B' > A$,

$$\left| \int_a^B f(x)dx \int_\alpha^{B'} h(y)e^{i(\beta x+t)y}dG(y) - \int_a^b f(x)dx \int_\alpha^{b'} h(y)e^{i(\beta x+t)y}dG(y) \right|$$

$$\leqq \left| \int_b^B |f(x)|\,dx \int_\alpha^{B'} |h(y)|\,dG(y) \right| + \left| \int_a^B |f(x)|\,dx \int_{b'}^{B'} |h(y)|\,dG(y) \right| < \varepsilon \ .$$

Hence by the Cauchy criterion for real numbers our theorem is proved.

Combining this result with Theorem 4.2 leads to:

Theorem 4.4. If f is continuous and absolutely integrable on $[a, \infty)$, and if h is continuous and bounded on $[\alpha, \infty)$, and if G is bounded and monotone increasing on $[\alpha, \infty)$, then for real β and t,

$$\int_a^\infty f(x)\,dx \int_\alpha^\infty h(y)\,e^{i(\beta x+t)y}dG(y)$$

and

$$\int_\alpha^\infty h(y)\,dG(y) \int_a^\infty f(x)e^{i(\beta x+t)y}dx$$

both exist and are equal.

Corresponding to Corollary 4.1 we have the expected extensions of Theorems 4.2, 4.3, and 4.4 to three variables. For example, from Theorem 4.4 we deduce:

Corollary 4.2. If f is continuous and absolutely integrable on $[a, \infty)$, and if g is continuous and absolutely integrable on $[c, \infty)$, and if h is continuous and bounded on $[\alpha,\infty)$, and if G is bounded and monotone increasing on $[\alpha, \infty)$, then for real β, γ, t the six iterated integrals of

$$f(x)g(y)h(\lambda)\,e^{i(\beta x+\gamma y+t)\lambda}$$

on $[a, \infty) \times [c, \infty) \times [\alpha, \infty)$ with respect to x, y, $G(\lambda)$, all exist and are equal.

5. Input–Output Covariances

Returning to (4.1), we see from Corollary 4.2 that (4.2) is valid if H is absolutely integrable on $[0, \infty)$. But this will be true if the zeros of the indicial polynomial Q of $\mathscr{L}$ all have negative real parts. Thus we conclude:

Theorem 5.1. Let $\{\xi(t) \mid -\infty < t < \infty\}$ be a mean zero wide-sense stationary complex stochastic process with continuous covariance function R and spectral distribution function S. Let H be the one-sided Green's function for the linear differential operator

$$\mathscr{L} = p_0\frac{d^n}{dt^n} + p_1\frac{d^{n-1}}{dt^{n-1}} + \cdots + p_n\,\cdot$$

where the p_j, $0 \leqq j \leqq n$, are complex constants, and $p_0 \neq 0$. Let all the zeros of the indicial polynomial

$$Q(\lambda) = p_0\lambda^n + p_1\lambda^{n-1} + \cdots + p_n$$

have negative real parts. Then the stochastic differential equation

$$\mathscr{L}z(t) = \xi(t)$$

defines a wide-sense stationary mean zero complex stochastic process $\{z(t)| -\infty < t < \infty\}$,

$$z(t) = \int_{-\infty}^{t} H(t-\tau)\xi(\tau)d\tau \tag{5.1}$$

with covariance function

$$r(h) = \int_{-\infty}^{\infty} |M(\lambda)|^2 e^{i2\pi h\lambda} dS(\lambda) \tag{5.2}$$

where

$$M(\lambda) = \int_{0}^{\infty} H(x)e^{-i2\pi\lambda x}dx \quad . \tag{5.3}$$

The function M is closely related to the indicial polynomial Q. For from

$$\mathscr{L}e^{i2\pi\lambda t} = Q(i2\pi\lambda)e^{i2\pi\lambda t}$$

follows

$$e^{i2\pi\lambda t} = \int_{0}^{\infty} H(x)[Q(i2\pi\lambda)e^{i2\pi\lambda(t-x)}]dx$$

or

$$1 = Q(i2\pi\lambda)\int_{0}^{\infty} H(x)e^{-i2\pi\lambda x}dx = Q(i2\pi\lambda)M(\lambda)$$

and

$$M(\lambda) = Q^{-1}(i2\pi\lambda) \quad .$$

If we let

$$\begin{aligned} h(x) &= H(x) \quad , \qquad x > 0 \; , \\ h(x) &= 0 \qquad\quad , \qquad x < 0 \; , \end{aligned} \tag{5.4}$$

then it is customary to call h the *impulsive response* of the *system* (operator) $\mathscr{L}^{-1}$(see [Mil 9]). In this case M is the Fourier transform of h and is called the *system function* of $\mathscr{L}^{-1}$.

Using the definition (5.4) of h, we may write (5.1) as

$$z(t) = \int_{-\infty}^{\infty} h(t-\tau)\xi(\tau)\,d\tau \quad , \tag{5.5}$$

from which (5.2) follows. Now one may start with (5.5), that is, assume that the two stochastic processes $\{\xi(t)| -\infty < t < \infty\}$ and $\{z(t)| -\infty < t <$

$\infty\}$ are related by the convolution integral (5.5). Then with the obvious conditions on $\{\xi(t) | -\infty < t < \infty\}$ and h (see Theorem 5.1) we may deduce the covariance function of the $\{z(t) | -\infty < t < \infty\}$ process in terms of the spectral distribution function of the $\{\xi(t) | -\infty < t < \infty\}$ process. The point we are making is that one can deduce (5.2) for any linear system with impulsive response h without explicitly mentioning the differential operator $\mathscr{L}$. In engineering practice one generally considers a "block diagram" consisting of a "black box" or "system" $\mathscr{S}$ with "input" $\xi(t)$ and "output" $z(t)$. The system is characterized by its impulsive response h or system function M (the Fourier transform of h).

Since r [cf. (5.2)] is a covariance function, and since it is certainly continuous, Bochner's theorem implies the existence of a spectral distribution function, say s, such that

$$r(h) = \int_{-\infty}^{\infty} e^{i2\pi h\lambda} ds(\lambda) \quad . \tag{5.6}$$

Now compare (5.6) and (5.2). We see, for example, that if s and S are absolutely continuous, then $s' = w$ and $S' = W$ a. e. (Lebesgue measure). By the uniqueness of the Fourier inversion theorem we have the well-known result that

$$w(\lambda) = |M(\lambda)|^2 W(\lambda) \tag{5.7}$$

except possibly on a set of Lebesgue measure zero. In words, the output spectral density is equal to the square of the modulus of the Fourier transform of the impulsive response of the system multiplied by the input spectral density.

Various generalizations of the formula above may be made. One trivial one related to (3.7) (see Corollary 3.1) is

Corollary 5.1. Let g be a complex-valued function of the real variable t continuous and bounded on $(-\infty, \infty)$. Then under the hypotheses of Theorem 5.1, the stochastic differential equation

$$\mathscr{L} y(t) = g(t)\xi(t)$$

defines a mean zero stochastic process $\{y(t) | -\infty < t < \infty\}$ with covariance function

$$c(t, \tau) = \int_{-\infty}^{\infty} N(\lambda, t)\bar{N}(\lambda, \tau) e^{i2\pi(t-\tau)\lambda} dS(\lambda) \tag{5.8}$$

where

$$N(\lambda, t) = \int_0^{\infty} H(x) g(t-x) e^{-i2\pi\lambda x} dx \quad .$$

If $g(t) \equiv 1$, then (5.8) reduces to (5.2).

6. Linear Difference Equations

Let $I = \{\cdots, -1, 0, +1, \cdots\}$ be the set of all integers. If $a \in I$, we define $I_a = \{a, a+1, \cdots\} \subset I$. Let p_j, $0 \leqq j \leqq n$, be complex-valued functions of the real variable t defined on I_a, and let $p_0(t)\, p_n(t) \neq 0$ on I_a. Then if $\{\xi(t) \mid t \in I_a\}$ is a random process, the equation

$$\mathscr{L} z(t) = p_0(t) z(t+n) + p_1(t) z(t+n-1) + \cdots + p_n(t) z(t) = \xi(t) \quad (6.1)$$

defines a new random process $\{z(t) \mid t \in I_a\}$. Equation (6.1) is a *stochastic difference equation*, and $\{z(t) \mid t \in I_a\}$ is called an *autoregressive process*. (See, for example, [Mil 6, 7].)

The similarity between difference equations and differential equations is well known. In this and the next section we shall deduce a few elementary properties of linear difference equations analogous to those proved in Sections 3 and 5 for linear differential equations.

First we recall that the homogeneous equation

$$\mathscr{L}\phi(t) = 0$$

has n linearly independent solutions, say $\phi_1, \cdots, \phi_n$ (see [Mil 5, page 40]). The one-sided Green's function H of $\mathscr{L}$ is defined as

$$H(t,s) = \frac{(-1)^{n-1}}{p_0(s) C(s+1)} \begin{vmatrix} \phi_1(t) & \phi_2(t) & \cdots & \phi_n(t) \\ \phi_1(s+1) & \phi_2(s+1) & \cdots & \phi_n(s+1) \\ \phi_1(s+2) & \phi_2(s+2) & \cdots & \phi_n(s+2) \\ \vdots & \vdots & & \vdots \\ \phi_1(s+n-1) & \phi_2(s+n-1) & \cdots & \phi_n(s+n-1) \end{vmatrix}$$

where

$$C(s) = \begin{vmatrix} \phi_1(s) & \phi_2(s) & \cdots & \phi_n(s) \\ \phi_1(s+1) & \phi_2(s+1) & \cdots & \phi_n(s+1) \\ \vdots & \vdots & & \vdots \\ \phi_1(s+n-1) & \phi_2(s+n-1) & \cdots & \phi_n(s+n-1) \end{vmatrix}$$

is the Casorati.

One may now readily verify that if ψ is a function with domain I_a, then

$$\phi(t) = \sum_{s=a}^{t-1} H(t,s) \psi(s) \quad (6.2)$$

is the unique solution of

$$\mathscr{L}\phi(t) = \psi(t)$$

which assumes the initial values $\phi(a+k)=0, 0 \leqq k \leqq n-1$. In (6.2) we are using the usual convention that the sum from a to b is vacuous if $b < a$.

Suppose now $\{\xi(t) \mid t \in I\}$ is a mean zero complex stochastic process with covariance function R. Then if

$$z_a(t) = \sum_{s=a}^{t-1} H(t, s)\xi(s) \quad , \tag{6.3}$$

the $\{z_a(t) \mid t \in I_a\}$ process is a solution of the difference equation

$$\mathscr{L} z_a(t) = \xi(t)$$

where $z_a(a+k) = 0, 0 \leqq k \leqq n-1$, with probability one, and has mean zero and covariance function

$$r_a(t, s) = \sum_{u=a}^{t-1} \sum_{v=a}^{s-1} H(t, u)\, \bar{H}(s, v)\, R(u, v) \quad . \tag{6.4}$$

Suppose now we let a decrease to negative infinity. Then by Theorem 6.4 of Chapter I,

$$z(t) = \sum_{s=-\infty}^{t-1} H(t, s)\xi(s)$$

will exist in mean square provided

$$\sum_{u=-\infty}^{t} |H(t, u)| (R(u, u))^{1/2} < \infty \tag{6.5}$$

for all t; and

$$r(t, s) = \sum_{u=-\infty}^{t-1} \sum_{v=-\infty}^{s-1} H(t, u)\bar{H}(s, v) R(u, v) \tag{6.6}$$

also exists.

Now let us make the additional assumption that $\mathscr{L}$ has *constant coefficients*, say

$$\mathscr{L}\phi(t) = p_0\phi(t+n) + p_1\phi(t+n-1) + \cdots + p_n\phi(t) \tag{6.7}$$

where the p_j, $0 \leqq j \leqq n$, are complex constants and $p_0 p_n \neq 0$. Then the Green's function H of $\mathscr{L}$ is a function only of the difference of its arguments, $H(t, s) = H(t-s)$, and (6.6) becomes

$$r(t, s) = \sum_{x=1}^{\infty} \sum_{y=1}^{\infty} H(x)\bar{H}(y) R(t-x, s-y) \quad . \tag{6.8}$$

Finally, if we make the further assumption that $\{\xi(t) \mid t \in I\}$ is wide-sense stationary, then the covariance function R of $\{\xi(t)\}$ is a function only of the difference of its arguments, $R(u, v) = R(u-v)$, and (6.8) becomes

$$r(t, s) = \sum_{x=1}^{\infty} \sum_{y=1}^{\infty} H(x)\bar{H}(y)\, R\,(t - s - x + y) \quad . \tag{6.9}$$

Note that $r(t,s)$ is now a function only of $t-s$. Hence the stochastic difference equation $\mathscr{L}z(t) = \xi(t)$ defines a wide-sense stationary autoregressive process $\{z(t) \mid t \in I\}$.

From (6.9),

$$|r(t - s)| \leqq R(0)\left[\sum_{x=0}^{\infty} |H(x)|\right]^2$$

and the condition (6.5) reduces to the condition that the series

$$\sum_{x=0}^{\infty} H(x) \tag{6.10}$$

be absolutely convergent. Furthermore, when $\mathscr{L}$ has constant coefficients, we see that if all the zeros of the indicial polynomial [see (6.7)]

$$Q(\lambda) = p_0\lambda^n + p_1\lambda^{n-1} + \cdots + p_n \tag{6.11}$$

are less than one in modulus, then the series of (6.10) is indeed absolutely convergent.

7. Autoregressive Processes

Let $\{\xi(t) | t \in I\}$ be a wide-sense stationary complex stochastic process with mean zero and covariance sequence $\{R(t) | t \in I\}$. Then there exists a unique monotone increasing function S, called the *spectral distribution function* (see [An, page 383]), such that

$$R(t) = \int_{-\frac{1}{2}}^{\frac{1}{2}} e^{i2\pi tf} dS(f) \quad , \qquad t \in I \quad . \tag{7.1}$$

This result is analogous to Bochner's theorem (Theorem 5.1 of Chapter II) for continuous parameter processes. Hence, if all the zeros of the indicial polynomial Q are less than one in modulus, we may write (6.9) as

$$r(h) = \sum_{x=1}^{\infty} \sum_{y=1}^{\infty} H(x)\bar{H}(y) R(h - x + y)$$

and from (7.1)

$$r(h) = \int_{-\frac{1}{2}}^{\frac{1}{2}} |M(\lambda)|^2 e^{i2\pi h\lambda} dS(\lambda) \tag{7.2}$$

where

$$M(\lambda) = \sum_{x=1}^{\infty} H(x) e^{-i2\pi\lambda x} \quad . \tag{7.3}$$

The function M is closely related to the indicial polynomial Q [see (6.11)]. In fact, we shall show that

$$M(\lambda) = Q^{-1}(e^{i2\pi\lambda}) \quad . \tag{7.4}$$

For, since

$$\mathscr{L}\phi(t) = \psi(t) \tag{7.5}$$

implies

$$\phi(t) = \sum_{s=1}^{\infty} H(s)\psi(t-s) \quad , \tag{7.6}$$

(provided ψ is bounded) we have only to let

$$\phi(t) = e^{i2\pi\lambda t} \quad .$$

Then from (7.5)

$$\psi(t) = Q(e^{i2\pi\lambda})\phi(t)$$

and from (7.6)

$$e^{i2\pi\lambda t} = \sum_{s=1}^{\infty} H(s)\, Q(e^{i2\pi\lambda})\, e^{i2\pi\lambda(t-s)}$$

or

$$\begin{aligned} 1 &= Q(e^{i2\pi\lambda}) \sum_{s=1}^{\infty} H(s) e^{-i2\pi\lambda s} \\ &= Q(e^{i2\pi\lambda})\, M(\lambda) \quad , \end{aligned}$$

which is (7.4).

We call M the *system function* of $\mathscr{L}^{-1}$ (see page 132). Now if $\mathscr{L}$ is a *differential* operator, the impulsive response of the system may be defined as the response to a delta function input. Thus [see (5.1)]

$$\begin{aligned} z(t) = \int_{-\infty}^{t} H(t-\tau)\delta(\tau)\,d\tau &= H(t) \quad , \qquad t > 0 \quad , \\ &= 0 \qquad\quad , \qquad t < 0 \quad . \end{aligned}$$

That is, the impulsive response is $h(t)$. Similarly we may define the impulsive response of the *difference* system as the response to $\psi(t) = \delta_{t\tau}$ where $\delta_{t\tau}$ is the Kronecker delta. Then [see (7.6)]

$$\phi(t) = \sum_{s=1}^{\infty} H(s)\delta_{t-s,\tau} = H(t-\tau) \quad , \qquad t > \tau \quad ,$$

$$= 0 \quad , \qquad t \leqq \tau \quad .$$

Let s be the spectral distribution function of the $\{z(t) | t \in I\}$ process (with covariance sequence $\{r(t) | t \in I\}$). Then

$$r(t) = \int_{-\frac{1}{2}}^{\frac{1}{2}} e^{i2\pi tf} ds(f) \quad , \qquad t \in I \quad . \tag{7.7}$$

Now compare (7.7) and (7.2). We see, for example, that if s and S are absolutely continuous, then $s' = w$ and $S' = W$ a.e. Thus, by the uniqueness of the spectral distribution function, we have

$$w(\lambda) = |Q(e^{i2\pi\lambda})|^{-2} W(\lambda) \quad . \tag{7.8}$$

In the case in which the spectral distribution function is absolutely continuous, we call its derivative the *spectral density.*

If $\mathscr{L}$ has constant coefficients and $\{\xi(t) | t \in I\}$ is wide-sense stationary, then from (6.3) or (6.4)

$$r_a(t, s) = \sum_{x=1}^{t-a} \sum_{y=1}^{s-a} H(x)\bar{H}(y)R(t - s - x + y) \quad .$$

Thus for finite a the $\{z_a(t) | t \in I_a\}$ process is not necessarily stationary. However, from (7.1) we may write

$$r_a(t, s) = \int_{-\frac{1}{2}}^{\frac{1}{2}} \left[\sum_{x=1}^{t-a} H(x)e^{-i2\pi fx}\right]\left[\sum_{y=1}^{s-a} \bar{H}(y)e^{i2\pi fy}\right] e^{i2\pi(t-s)f} dS(f) \quad .$$

Note that from

$$\mathscr{L}z(t) = \xi(t)$$

we may calculate

$$\mathscr{E}[\mathscr{L}z(t)]\overline{[\mathscr{L}z(s)]} = \mathscr{E}\xi(t)\bar{\xi}(s) = R(t-s)$$

as

$$R(h) = \sum_{\mu=0}^{n} \sum_{\nu=0}^{n} p_\mu \bar{p}_\nu r(h + \nu - \mu) \quad .$$

And using (7.7)

$$R(h) = \int_{-\frac{1}{2}}^{\frac{1}{2}} |Q(e^{i2\pi\lambda})|^2 e^{i2\pi h\lambda} ds(\lambda) \quad ,$$

which is an "inverse" of (7.2). Thus under the assumptions that s and S are absolutely continuous, we have a more direct proof of (7.8) .

V

ESTIMATION OF SIGNALS IN ADDITIVE NOISE

0. Introduction

Let $\{z(t)\}$ be a stochastic process depending on various parameters $\boldsymbol{\alpha}$. Based on a finite number of observations on $\{z(t)\}$ we would like to determine "good" estimates of the components of $\boldsymbol{\alpha}$. This is a basic statistical problem. We shall consider expressions of the form

$$z(t) = c(t) + x(t) + y(t)$$

where c is a deterministic (complex) signal of t (time) and x and y are members of independent complex Gaussian processes. We shall assume that c and x are also functions of the (real) vector parameter $\boldsymbol{\alpha}$, but that y is not. One frequently calls c and x the "signal" and y the "noise." Thus from a communication engineering point of view we are trying to estimate the signal in the presence of additive noise; while from a mathematical statistical viewpoint we are attempting to estimate the parameters of a distribution. Specifically, we shall determine the maximum likelihood equations and the Fisher information matrix for the problem of estimating the parameters of deterministic and random signals embedded in complex Gaussian noise. In Section 1 we illustrate the ideas involved with a simple example.

Section 2 is devoted to a number of propaedeutic results that are precursor to the succeeding analyses, while Section 3 is concerned with the general formulation of the likelihood equations and the Cramér–Rao lower bounds. The stochastic model is introduced in Section 4, and explicit formulas are determined for the general case. Finally, in Sections 5 and 6 we discuss some nontrivial practical problems of general interest.

1. An Elementary Example

Let

$$z(t) = \alpha c(t) + y(t) \tag{1.1}$$

where c is a known complex signal defined on $(-\infty, \infty)$ and $\{y(t) \mid -\infty < t < \infty\}$ is a mean zero complex stationary Gaussian process with positive definite covariance function R (and independent of α). Let $t_1, \cdots, t_m$ be any set of m distinct time points such that not all the $c(t_k)$, $1 \leqq k \leqq m$, are zero, and let

$$z(t_1), \cdots, z(t_m)$$

be m independent observations on $\{z(t) \mid -\infty < t < \infty\}$. From these data we shall construct an unbiased efficient estimator of the parameter α. From a statistical point of view we are estimating the mean of a Gaussian distribution; from the communication viewpoint we are estimating the amplitude of a deterministic signal immersed in additive Gaussian noise.

Referring to (1.1) we see that z is a complex Gaussian variate with mean $\mathscr{E} z = \alpha c$ and variance $\operatorname{Var} z = \mathscr{E} |z - \alpha c|^2 = R(0) > 0$. The density function of z is (see Chapter III)

$$f(z; \alpha) = \frac{1}{\pi R(0)} \exp\left(\frac{-|z - \alpha c|^2}{R(0)}\right) .$$

For convenience we write $z_k = z(t_k)$ and $c_k = c(t_k)$ for $1 \leqq k \leqq m$. Then the likelihood function l of the m observations $z_1, \cdots, z_m$ is

$$l(z_1, \cdots, z_m; \alpha) = \frac{1}{\pi^m R^m(0)} \prod_{k=1}^{m} \exp\left(\frac{-|z_k - \alpha c_k|^2}{R(0)}\right) .$$

A nonconstant solution of the likelihood equation

$$\frac{\partial}{\partial \alpha} \log l = 0$$

yields a maximum likelihood estimator $\hat{\alpha}$ of α. Now

$$\frac{\partial}{\partial \alpha} \log l = \frac{1}{R(0)} \sum_{k=1}^{m} [(z_k - \alpha c_k)\bar{c}_k + (\bar{z}_k - \alpha \bar{c}_k) c_k] \tag{1.2}$$

and hence

$$\hat{\alpha} = \frac{\sum_{k=1}^{m} \operatorname{Re}[c_k \bar{z}_k]}{\sum_{k=1}^{m} |c_k|^2}. \tag{1.3}$$

From Chapter III, or by a direct calculation,

$$\mathscr{E} \hat{\alpha} = \alpha .$$

Thus $\hat{\alpha}$ is an unbiased estimator of α.

The Cramér–Rao lower bound C for the variance of any unbiased estimator of α is given by

$$C = \frac{1}{\mathscr{E}\left(\frac{\partial}{\partial\alpha}\log l\right)^2} .$$

From (1.3) we have, by the independence of the z_k observations, that

$$\begin{aligned}
\mathscr{E}\left(\frac{\partial}{\partial\alpha}\log l\right)^2 &= \frac{1}{R^2(0)}\sum_{k=1}^{m}\mathscr{E}[(z_k - \alpha c_k)\bar{c}_k + (\bar{z}_k - \alpha\bar{c}_k)c_k]^2 \\
&= \frac{2}{R^2(0)}\sum_{k=1}^{m}|c_k|^2\mathscr{E}|z_k - \alpha c_k|^2 \\
&= \frac{2}{R(0)}\sum_{k=1}^{m}|c_k|^2
\end{aligned}$$

and hence

$$C = \frac{R(0)}{2\sum_{k=1}^{m}|c_k|^2} .$$

But from (1.3)

$$\operatorname{Var}\hat{\alpha} = \frac{\sum_{k=1}^{m}|c_k|^2 \operatorname{Var} z_k}{2\left(\sum_{k=1}^{m}|c_k|^2\right)^2} = \frac{R(0)}{2\sum_{k=1}^{m}|c_k|^2} .$$

Thus since $\operatorname{Var}\hat{\alpha} = C$, we infer that $\hat{\alpha}$ is an efficient statistic (and, *a fortiori*, a sufficient statistic) for estimating α.

2. Some Preliminary Results

In order to efficiently formulate our estimation problems in the general case we need a few mathematical results from the theory of complex multivariate analysis. These formulas are derived in Lemmas 2.1 and 2.2, and in Theorem 2.1.

Lemma 2.1. Let α be a real variable. Let $\boldsymbol{P} = \boldsymbol{P}(\alpha)$ be a continuously differentiable nonsingular Hermitian matrix for all $\alpha \in \operatorname{Dom}\boldsymbol{P}$. Then

$$(\det \boldsymbol{P})^{-1}\frac{\partial}{\partial\alpha}\det \boldsymbol{P} = \operatorname{tr}\boldsymbol{P}^{-1}\frac{\partial \boldsymbol{P}}{\partial\alpha} .$$

Proof. Since $\boldsymbol{P}$ is a Hermitian matrix, there exists a unitary matrix $\boldsymbol{U}$ such that

$$\bar{\boldsymbol{U}}'\boldsymbol{P}\boldsymbol{U} = \boldsymbol{D}$$

where $\boldsymbol{D}$ is a diagonal matrix whose diagonal elements are the real, but not necessarily distinct, characteristic zeros of $\boldsymbol{P}$. Thus,

$$
\begin{aligned}
(\det \boldsymbol{P})^{-1} \frac{\partial}{\partial \alpha} \det \boldsymbol{P} &= \operatorname{tr} \boldsymbol{D}^{-1} \frac{\partial \boldsymbol{D}}{\partial \alpha} \\
&= \operatorname{tr}(\bar{\boldsymbol{U}}'\boldsymbol{P}^{-1}\boldsymbol{U})\left(\frac{\partial \bar{\boldsymbol{U}}'}{\partial \alpha}\boldsymbol{P}\boldsymbol{U} + \bar{\boldsymbol{U}}'\frac{\partial \boldsymbol{P}}{\partial \alpha}\boldsymbol{U} + \bar{\boldsymbol{U}}'\boldsymbol{P}\frac{\partial \boldsymbol{U}}{\partial \alpha}\right) \\
&= \operatorname{tr}\left(\boldsymbol{U}\frac{\partial \bar{\boldsymbol{U}}'}{\partial \alpha} + \boldsymbol{P}^{-1}\frac{\partial \boldsymbol{P}}{\partial \alpha} + \frac{\partial \boldsymbol{U}}{\partial \alpha}\bar{\boldsymbol{U}}'\right) \\
&= \operatorname{tr}\left(\frac{\partial}{\partial \alpha}\boldsymbol{U}\bar{\boldsymbol{U}}' + \boldsymbol{P}^{-1}\frac{\partial \boldsymbol{P}}{\partial \alpha}\right) \\
&= \operatorname{tr} \boldsymbol{P}^{-1}\frac{\partial \boldsymbol{P}}{\partial \alpha}
\end{aligned}
$$

since $\boldsymbol{U}\bar{\boldsymbol{U}}' = \boldsymbol{I}$, the identity matrix.

Lemma 2.2. Let $\boldsymbol{z}$ be normally distributed $(\boldsymbol{0}, \boldsymbol{R})$ in n complex dimensions, and let $\boldsymbol{R}$ be positive definite. Let $\boldsymbol{H}$ and $\boldsymbol{K}$ be any $n \times n$ Hermitian matrices. Then

$$\mathscr{E}\bar{\boldsymbol{z}}'\boldsymbol{H}\boldsymbol{z} = \operatorname{tr} \boldsymbol{R}\boldsymbol{H}$$

and

$$\mathscr{E}(\bar{\boldsymbol{z}}'\boldsymbol{H}\boldsymbol{z})(\bar{\boldsymbol{z}}'\boldsymbol{K}\boldsymbol{z}) = (\operatorname{tr} \boldsymbol{R}\boldsymbol{H})(\operatorname{tr} \boldsymbol{R}\boldsymbol{K}) + \operatorname{tr} \boldsymbol{R}\boldsymbol{H}\boldsymbol{R}\boldsymbol{K} \quad .$$

Proof. Since $\boldsymbol{R}$ is positive definite, there exists a nonsingular matrix $\boldsymbol{W}$ such that

$$\boldsymbol{R} = \bar{\boldsymbol{W}}'\boldsymbol{W} \quad .$$

Let $\boldsymbol{\xi} = \bar{\boldsymbol{W}}'^{-1}\boldsymbol{z}$. Then

$$\mathscr{E}\boldsymbol{\xi}\bar{\boldsymbol{\xi}}' = \bar{\boldsymbol{W}}'^{-1}(\mathscr{E}\boldsymbol{z}\bar{\boldsymbol{z}}')\boldsymbol{W}^{-1} = \bar{\boldsymbol{W}}'^{-1}\boldsymbol{R}\boldsymbol{W}^{-1} = \boldsymbol{I}$$

where $\boldsymbol{I}$ is the $n \times n$ identity matrix. Thus $\boldsymbol{\xi}$ is normal $(\boldsymbol{0}, \boldsymbol{I})$.

Now

$$\mathscr{E}\bar{\boldsymbol{z}}'\boldsymbol{H}\boldsymbol{z} = \mathscr{E}\bar{\boldsymbol{\xi}}'\boldsymbol{W}\boldsymbol{H}\bar{\boldsymbol{W}}'\boldsymbol{\xi} = \mathscr{E}\bar{\boldsymbol{\xi}}'\boldsymbol{P}\boldsymbol{\xi}$$

where $\boldsymbol{P} = \boldsymbol{W}\boldsymbol{H}\bar{\boldsymbol{W}}' = \| P_{jk} \|$. Then if we write

$$\boldsymbol{\xi}' = \{\xi_1, \cdots, \xi_n\} \quad ,$$

we have

$$\begin{aligned}\mathscr{E}\bar{\boldsymbol{z}}'\boldsymbol{H}\boldsymbol{z} &= \mathscr{E}\sum_{j=1}^{n}\sum_{k=1}^{n} P_{jk}\bar{\xi}_j\xi_k \\ &= \mathscr{E}\sum_{j} P_{jj}|\xi_j|^2 + \mathscr{E}\sum_{\substack{j,\,k\\ j\neq k}} P_{jk}\bar{\xi}_j\xi_k \\ &= \sum_{j} P_{jj} = \operatorname{tr}\boldsymbol{P} \quad .\end{aligned}$$

But

$$\operatorname{tr}\boldsymbol{P} = \operatorname{tr}\boldsymbol{W}\boldsymbol{H}\bar{\boldsymbol{W}}' = \operatorname{tr}\bar{\boldsymbol{W}}'\boldsymbol{W}\boldsymbol{H} = \operatorname{tr}\boldsymbol{R}\boldsymbol{H} \quad ,$$

and the first part of the lemma is proved.

Similarly

$$\mathscr{E}(\bar{\boldsymbol{z}}'\boldsymbol{H}\boldsymbol{z})\,(\bar{\boldsymbol{z}}'\boldsymbol{K}\boldsymbol{z}) = \mathscr{E}(\bar{\boldsymbol{\xi}}'\boldsymbol{P}\boldsymbol{\xi})\,(\bar{\boldsymbol{\xi}}'\boldsymbol{Q}\boldsymbol{\xi})$$

where $\boldsymbol{P}$ has been defined above and

$$\boldsymbol{Q} = \boldsymbol{W}\boldsymbol{K}\bar{\boldsymbol{W}}' = \|\,Q_{jk}\,\| \quad .$$

Then

$$\begin{aligned}\mathscr{E}(\bar{\boldsymbol{z}}'\boldsymbol{H}\boldsymbol{z})\,(\bar{\boldsymbol{z}}'\boldsymbol{K}\boldsymbol{z}) &= \mathscr{E}\Big(\sum_{j,\,k} P_{jk}\bar{\xi}_j\xi_k\Big)\Big(\sum_{\alpha,\beta} Q_{\alpha\beta}\bar{\xi}_\alpha\xi_\beta\Big) \\ &= \mathscr{E}\Big(\sum_{j} P_{jj}|\xi_j|^2\Big)\Big(\sum_{\alpha} Q_{\alpha\alpha}|\xi_\alpha|^2\Big) \\ &\quad + \mathscr{E}\Big(\sum_{j\neq k} P_{jk}\bar{\xi}_j\xi_k\Big)\Big(\sum_{\alpha\neq\beta} Q_{\alpha\beta}\bar{\xi}_\alpha\xi_\beta\Big) \\ &= \mathscr{E}\sum_{j} P_{jj}Q_{jj}|\xi_j|^4 + \mathscr{E}\sum_{j\neq\alpha} P_{jj}Q_{\alpha\alpha}|\xi_j|^2|\xi_\alpha|^2 \\ &\quad + \mathscr{E}\sum_{j\neq k} P_{jk}Q_{kj}|\xi_j|^2|\xi_k|^2 \\ &= 2\sum_{j} P_{jj}Q_{jj} + \sum_{j\neq\alpha} P_{jj}Q_{\alpha\alpha} + \sum_{j\neq k} P_{jk}Q_{kj} \\ &= \sum_{j,\alpha} P_{jj}Q_{\alpha\alpha} + \sum_{j,k} P_{jk}Q_{kj} \\ &= (\operatorname{tr}\boldsymbol{P})\,(\operatorname{tr}\boldsymbol{Q}) + \operatorname{tr}\boldsymbol{P}\boldsymbol{Q} \\ &= (\operatorname{tr}\boldsymbol{R}\boldsymbol{H})\,(\operatorname{tr}\boldsymbol{R}\boldsymbol{K}) + \operatorname{tr}\boldsymbol{R}\boldsymbol{H}\boldsymbol{R}\boldsymbol{K}\end{aligned}$$

—which proves the second part of the lemma.

Finally, we shall need:

Theorem 2.1. Let α and β be real parameters. Let $\boldsymbol{R} = \boldsymbol{R}(\alpha, \beta)$ be a continuously differentiable positive definite Hermitian matrix for all $(\alpha, \beta) \in \operatorname{Dom}\boldsymbol{R}$. Let $\boldsymbol{c} = \boldsymbol{c}(\alpha, \beta)$ be a continuously differentiable complex-valued vector for all $(\alpha, \beta) \in \operatorname{Dom}\boldsymbol{c}$. Let $\boldsymbol{z}$ be normally distributed $(\boldsymbol{c}, \boldsymbol{R})$ in n complex dimensions. Let f be the complex density function of $\boldsymbol{z}$. Then for all $(\alpha, \beta) \in (\operatorname{Dom}\boldsymbol{R}) \cap (\operatorname{Dom}\boldsymbol{c})$,

$$\mathscr{E}\left(\frac{\partial}{\partial\alpha}\log f\right)\left(\frac{\partial}{\partial\beta}\log f\right) = \operatorname{tr} \boldsymbol{R}\frac{\partial \boldsymbol{S}}{\partial\alpha}\boldsymbol{R}\frac{\partial \boldsymbol{S}}{\partial\beta} + 2\operatorname{Re}\frac{\partial \bar{\boldsymbol{c}}'}{\partial\alpha}\boldsymbol{S}\frac{\partial \boldsymbol{c}}{\partial\beta}$$

where $\boldsymbol{S} = \boldsymbol{R}^{-1}$.

Proof. Explicitly, the density function of $\boldsymbol{z}$ is (see Chapter III)

$$f(\boldsymbol{z}) = \frac{1}{\pi^n \det \boldsymbol{R}} \exp\left[-(\bar{\boldsymbol{z}} - \bar{\boldsymbol{c}})'\boldsymbol{R}^{-1}(\boldsymbol{z} - \boldsymbol{c})\right] \quad ,$$

and

$$\frac{\partial}{\partial\alpha}\log f = (\det \boldsymbol{S})^{-1}\frac{\partial}{\partial\alpha}\det \boldsymbol{S} - (\bar{\boldsymbol{z}} - \bar{\boldsymbol{c}})'\frac{\partial \boldsymbol{S}}{\partial\alpha}(\boldsymbol{z} - \boldsymbol{c}) + \frac{\partial \bar{\boldsymbol{c}}'}{\partial\alpha}\boldsymbol{S}(\boldsymbol{z} - \boldsymbol{c}) + (\bar{\boldsymbol{z}} - \bar{\boldsymbol{c}})'\boldsymbol{S}\frac{\partial \boldsymbol{c}}{\partial\alpha}$$

where we have written $\boldsymbol{S} = \boldsymbol{R}^{-1}$. Thus

$$\begin{aligned}
&\mathscr{E}\left(\frac{\partial}{\partial\alpha}\log f\right)\left(\frac{\partial}{\partial\beta}\log f\right) \\
&= \mathscr{E}\left[(\det \boldsymbol{S})^{-1}\frac{\partial}{\partial\alpha}\det \boldsymbol{S} - \bar{\boldsymbol{\xi}}'\frac{\partial \boldsymbol{S}}{\partial\alpha}\boldsymbol{\xi} + \frac{\partial \bar{\boldsymbol{c}}'}{\partial\alpha}\boldsymbol{S}\boldsymbol{\xi} + \bar{\boldsymbol{\xi}}'\boldsymbol{S}\frac{\partial \boldsymbol{c}}{\partial\alpha}\right] \\
&\quad \times \left[(\det \boldsymbol{S})^{-1}\frac{\partial}{\partial\beta}\det \boldsymbol{S} - \bar{\boldsymbol{\xi}}'\frac{\partial \boldsymbol{S}}{\partial\beta}\boldsymbol{\xi} + \frac{\partial \bar{\boldsymbol{c}}'}{\partial\beta}\boldsymbol{S}\boldsymbol{\xi} + \bar{\boldsymbol{\xi}}'\boldsymbol{S}\frac{\partial \boldsymbol{c}}{\partial\beta}\right]
\end{aligned}$$

where

$$\boldsymbol{\xi} = \boldsymbol{z} - \boldsymbol{c}$$

has mean zero.

Hence by Lemmas 2.1 and 2.2,

$$\mathscr{E}\left(\frac{\partial}{\partial\alpha}\log f\right)\left(\frac{\partial}{\partial\beta}\log f\right) = \operatorname{tr} \boldsymbol{R}\frac{\partial \boldsymbol{S}}{\partial\alpha}\boldsymbol{R}\frac{\partial \boldsymbol{S}}{\partial\beta} + \mathscr{E}\frac{\partial \bar{\boldsymbol{c}}'}{\partial\alpha}\boldsymbol{S}\boldsymbol{\xi}\bar{\boldsymbol{\xi}}'\boldsymbol{S}\frac{\partial \boldsymbol{c}}{\partial\beta} + \mathscr{E}\bar{\boldsymbol{\xi}}'\boldsymbol{S}\frac{\partial \boldsymbol{c}}{\partial\alpha}\frac{\partial \bar{\boldsymbol{c}}'}{\partial\beta}\boldsymbol{S}\boldsymbol{\xi} \quad .$$

But $\mathscr{E}\boldsymbol{\xi}\bar{\boldsymbol{\xi}}' = \boldsymbol{R}$. Thus

$$\mathscr{E}\left(\frac{\partial}{\partial\alpha}\log f\right)\left(\frac{\partial}{\partial\beta}\log f\right) = \operatorname{tr} \boldsymbol{R}\frac{\partial \boldsymbol{S}}{\partial\alpha}\boldsymbol{R}\frac{\partial \boldsymbol{S}}{\partial\beta} + \frac{\partial \bar{\boldsymbol{c}}'}{\partial\alpha}\boldsymbol{S}\frac{\partial \boldsymbol{c}}{\partial\beta} + \frac{\partial \bar{\boldsymbol{c}}'}{\partial\beta}\boldsymbol{S}\frac{\partial \boldsymbol{c}}{\partial\alpha} \quad ,$$

and our theorem is proved.

Note that $\operatorname{tr} \boldsymbol{R}\frac{\partial \boldsymbol{S}}{\partial\alpha}\boldsymbol{R}\frac{\partial \boldsymbol{S}}{\partial\beta}$ is real since

$$\overline{\operatorname{tr} \boldsymbol{R}\frac{\partial \boldsymbol{S}}{\partial \alpha}\boldsymbol{R}\frac{\partial \boldsymbol{S}}{\partial \beta}} = \operatorname{tr} \bar{\boldsymbol{R}}\frac{\partial \bar{\boldsymbol{S}}}{\partial \alpha}\bar{\boldsymbol{R}}\frac{\partial \bar{\boldsymbol{S}}}{\partial \beta} = \operatorname{tr} \boldsymbol{R}'\frac{\partial \boldsymbol{S}'}{\partial \alpha}\boldsymbol{R}'\frac{\partial \boldsymbol{S}'}{\partial \beta}$$

$$= \operatorname{tr} \frac{\partial \boldsymbol{S}}{\partial \beta}\boldsymbol{R}\frac{\partial \boldsymbol{S}}{\partial \alpha}\boldsymbol{R} = \operatorname{tr} \boldsymbol{R}\frac{\partial \boldsymbol{S}}{\partial \alpha}\boldsymbol{R}\frac{\partial \boldsymbol{S}}{\partial \beta} \quad .$$

3. The Estimation Problem

Let $\boldsymbol{z}_1, \cdots, \boldsymbol{z}_m$ be m vector observations from an n-dimensional probability density function $f(\boldsymbol{z}; \boldsymbol{a})$ where $\boldsymbol{a}' = \{\alpha_1, \cdots, \alpha_p\}$ is a p-dimensional real parameter vector. Let $f(\boldsymbol{z}; \boldsymbol{a})$ be a continuously differentiable function of the α_j, $1 \leqq j \leqq p$, and let

$$L(\boldsymbol{z}_1, \cdots, \boldsymbol{z}_m; \boldsymbol{a}) = \prod_{k=1}^{m} f(\boldsymbol{z}_k; \boldsymbol{a})$$

be the likelihood function. The joint maximum likelihood estimates of the components of $\boldsymbol{a}$ are given by nonconstant solutions of the likelihood equations

$$\frac{\partial}{\partial \alpha_j} \log L = 0 \quad , \qquad 1 \leqq j \leqq p \quad .$$

Let

$$\boldsymbol{F} = \| F_{jk} \|$$

where

$$F_{jk} = \mathscr{E}\left(\frac{\partial}{\partial \alpha_j} \log f\right)\left(\frac{\partial}{\partial \alpha_k} \log f\right) \quad , \qquad 1 \leqq j, k \leqq p \quad ,$$

be the Fisher information matrix. Then

$$\operatorname{Var} \tilde{\alpha}_j \geqq \frac{1}{m} C_{jj} \quad , \qquad 1 \leqq j \leqq p \quad ,$$

where $\boldsymbol{F}^{-1} = \| C_{jk} \|$ and $\tilde{\alpha}_j = \tilde{\alpha}_j(\boldsymbol{z}_1, \cdots, \boldsymbol{z}_m)$ is any unbiased estimate of α_j.

Thus we have a standard procedure for computing estimators of parameters and a (mean square) criterion for determining their efficiency.

We remark in passing that if f has a continuous second partial derivative with respect to α_j and α_k, then

$$F_{jk} = -\mathscr{E}\frac{\partial^2 \log f}{\partial \alpha_j \partial \alpha_k} \quad .$$

For, by definition,

$$F_{jk} = \int_{-\infty}^{\infty}\int_{-\infty}^{\infty} f\left(\frac{\partial}{\partial\alpha_j}\log f\right)\left(\frac{\partial}{\partial\alpha_k}\log f\right)dx\, dy \quad .$$

Now consider the identity

$$\begin{aligned} f\left(\frac{\partial}{\partial\alpha_j}\log f\right)\left(\frac{\partial}{\partial\alpha_k}\log f\right) &= \frac{1}{f}\frac{\partial f}{\partial\alpha_j}\frac{\partial f}{\partial\alpha_k} \\ &= \frac{\partial^2 f}{\partial\alpha_j\partial\alpha_k} - f\frac{\partial^2 \log f}{\partial\alpha_j\partial\alpha_k} \end{aligned}$$

and note that

$$\int_{-\infty}^{\infty}\int_{-\infty}^{\infty}\frac{\partial^2 f}{\partial\alpha_j\partial\alpha_k}dx\, dy = \frac{\partial^2}{\partial\alpha_j\partial\alpha_k}\int_{-\infty}^{\infty}\int_{-\infty}^{\infty} f(\mathbf{z};\, \mathbf{a})dx\, dy = 0 \quad .$$

4. The Stochastic Model

Let $\{x(t) \mid -\infty < t < \infty\}$ and $\{y(t) \mid -\infty < t < \infty\}$ be independent mean zero complex Gaussian processes. Let $c(t)$ be a complex-valued deterministic function of t with domain $(-\infty, \infty)$, and let

$$z(t) = c(t) + x(t) + y(t) \quad . \tag{4.1}$$

Let $t_1, \cdots, t_n$ be n distinct time points and consider the vectors

$$\begin{aligned} \mathbf{z}' &= \{z(t_1), \cdots, z(t_n)\} \\ \mathbf{c}' &= \{c(t_1), \cdots, c(t_n)\} \\ \mathbf{x}' &= \{x(t_1), \cdots, x(t_n)\} \\ \mathbf{y}' &= \{y(t_1), \cdots, y(t_n)\} \quad . \end{aligned}$$

If $\mathbf{x}$ is normally distributed $(\mathbf{0}, \mathbf{R}_x)$ and if $\mathbf{y}$ is normally distributed $(\mathbf{0}, \mathbf{R}_y)$, then $\mathbf{z}$ is normal $(\mathbf{c}, \mathbf{R})$ where $\mathbf{R} = \mathbf{R}_x + \mathbf{R}_y$.

Let $\mathbf{a}$ be a real p-dimensional parameter vector defined on some p-dimensional region $\mathscr{A}$ of p-dimensional Euclidean space. We shall assume that $\mathbf{R}_x = \mathbf{R}_x(\mathbf{a})$ and $\mathbf{c} = \mathbf{c}(\mathbf{a})$ are continuously differentiable functions of the components of $\mathbf{a}$ for all $\mathbf{a}$ in $\mathscr{A}$. Also, we shall assume that $\mathbf{R}_y$ is independent of $\mathbf{a}$. Furthermore, we require that $\mathbf{R}_y$ be positive definite, and that $\mathbf{R}_x(\mathbf{a})$ be positive definite for all $\mathbf{a} \in \mathscr{A}$. We call $c(t)$ the deterministic signal, $x(t)$ the random signal process, and $y(t)$ the noise process. Our problem is to estimate the components of $\mathbf{a}$.

Now let $\{z_k(t) \mid -\infty < t < \infty\}$, $1 \leqq k \leqq m$, be m replications of the stochastic process $\{z(t) \mid -\infty < t < \infty\}$, and let

$$\mathbf{z}_k' = \{z_k(t_1), \cdots, z_k(t_n)\} \quad , \qquad 1 \leqq k \leqq m \quad , \tag{4.2}$$

be n observations on the stochastic process $\{z_k(t) \mid -\infty < t < \infty\}$. Thus we may consider $\boldsymbol{z}_1, \cdots, \boldsymbol{z}_m$ as vector observations from the n-dimensional probability density function

$$f(\boldsymbol{z}; \boldsymbol{a}) = \frac{1}{\pi^n \det \boldsymbol{R}(\boldsymbol{a})} \exp \{-[\bar{\boldsymbol{z}} - \bar{\boldsymbol{c}}(\boldsymbol{a})]' \boldsymbol{R}^{-1}(\boldsymbol{a}) [\boldsymbol{z} - \boldsymbol{c}(\boldsymbol{a})]\}$$

where

$$\boldsymbol{R}(\boldsymbol{a}) = \boldsymbol{R}_x(\boldsymbol{a}) + \boldsymbol{R}_y \quad .$$

We recall that if $\boldsymbol{\xi}$ is an n-dimensional vector and $\boldsymbol{A}$ is an $n \times n$ matrix, then we may write the scalar $q = \bar{\boldsymbol{\xi}}' \boldsymbol{A} \boldsymbol{\xi}$ as

$$q = \operatorname{tr} q = \operatorname{tr} \bar{\boldsymbol{\xi}}' \boldsymbol{A} \boldsymbol{\xi} = \operatorname{tr} \boldsymbol{A} \boldsymbol{\xi} \bar{\boldsymbol{\xi}}' \quad . \tag{4.3}$$

Thus, from Section 3, Lemma 2.1, and Theorem 2.1, the likelihood equations are

$$\operatorname{tr} \boldsymbol{R} \frac{\partial \boldsymbol{S}}{\partial \alpha_j} = \frac{1}{m} \sum_{k=1}^{m} \frac{\partial}{\partial \alpha_j} \operatorname{tr} \boldsymbol{S}(\boldsymbol{z}_k - \boldsymbol{c})(\bar{\boldsymbol{z}}_k - \bar{\boldsymbol{c}})' \quad , \qquad 1 \leqq j \leqq p \quad , \tag{4.4}$$

where $\boldsymbol{S} = \boldsymbol{R}^{-1}$, while the elements of the Fisher information matrix are

$$F_{jk} = \operatorname{tr} \boldsymbol{R} \frac{\partial \boldsymbol{S}}{\partial \alpha_j} \boldsymbol{R} \frac{\partial \boldsymbol{S}}{\partial \alpha_k} + 2 \operatorname{Re} \frac{\partial \bar{\boldsymbol{c}}'}{\partial \alpha_j} \boldsymbol{S} \frac{\partial \boldsymbol{c}}{\partial \alpha_k} \quad , \qquad 1 \leqq j, k \leqq p \quad . \tag{4.5}$$

[Incidentally, we could have used (4.3) to simplify the first part of the proof of Lemma 2.2; but, unfortunately, it does not seem to help in the proof of the second part.]

The computation of the elements of the inverse Fisher information matrix is generally direct, although frequently laborious. The main difficulty in any concrete problem stems from our ability (or inability) to solve the likelihood equations (4.4). Therefore, before proceeding to specific examples, let us examine in some detail the problem of solving the likelihood equations.

If we introduce the vector statistics

$$\tilde{\boldsymbol{R}} = \frac{1}{m} \sum_{k=1}^{m} (\boldsymbol{z}_k - \boldsymbol{c})(\bar{\boldsymbol{z}}_k - \bar{\boldsymbol{c}})' \tag{4.6}$$

and

$$\tilde{\boldsymbol{Z}} = \frac{1}{m} \sum_{k=1}^{m} \boldsymbol{z}_k \tag{4.7}$$

then we may write the likelihood equations of (4.4) in the compact form

$$\operatorname{tr} (\boldsymbol{R} - \tilde{\boldsymbol{R}}) \frac{\partial \boldsymbol{S}}{\partial \alpha_j} + 2 \operatorname{Re} \frac{\partial \bar{\boldsymbol{c}}'}{\partial \alpha_j} \boldsymbol{S}(\tilde{\boldsymbol{Z}} - \boldsymbol{c}) = 0 \quad , \qquad 1 \leqq j \leqq p. \tag{4.8}$$

Since the trace of the product of two $n \times n$ Hermitian matrices is real, (4.8) represents p real equations on the p unknowns $\alpha_1, \cdots, \alpha_p$. In general, in theory, one can solve these p equations for the α_j, $1 \leqq j \leqq p$. However, it is by no means obvious that this is always the case. For example, (4.8) may have no real solutions, and if it does have real solutions, they need not be unique. This last remark need not unduly concern us, for we know that any nontrivial solution of the likelihood equations yields a maximum likelihood estimator.

One "obvious" solution of (4.8) is to let

$$\boldsymbol{R}(\boldsymbol{a}) = \tilde{\boldsymbol{R}} \tag{4.9}$$

and

$$\boldsymbol{c}(\boldsymbol{a}) = \tilde{\boldsymbol{Z}} \tag{4.10}$$

—provided these equations can be satisfied simultaneously by some p-tuple $\boldsymbol{a}^0 = (\alpha_1^0, \cdots, \alpha_p^0)$. Since $\boldsymbol{R}(\boldsymbol{a})$ is an $n \times n$ Hermitian matrix, (4.9) is equivalent to n^2 real equations; while (4.10) is equivalent to $2n$ real equations. Thus if (4.9) and (4.10) are to be solutions of (4.8), it is reasonable to require that

$$n^2 + 2n = p \quad .$$

If $\boldsymbol{c}(\boldsymbol{a})$ is independent of $\boldsymbol{a}$, then we require that

$$n^2 = p \quad ,$$

and if $\boldsymbol{R}(\boldsymbol{a})$ is independent of $\boldsymbol{a}$, it is sufficient to require that

$$2n = p \quad .$$

Certain special cases are worthy of note. Suppose $\boldsymbol{c}(\boldsymbol{a})$ is absent and $\{z(t)\}$ is wide-sense stationary. Then if we write $\boldsymbol{R} = \| R_{jk} \|$, we see that the element R_{jk} of $\boldsymbol{R}$ depends only on the difference $j - k$. Thus we may satisfy (4.8) by letting

$$\tilde{R}(0) \equiv R_{ll} = \frac{1}{n} \sum_{i=1}^{n} \tilde{R}_{ii}$$

$$\tilde{R}(t_i - t_{i+\lambda}) \equiv R_{l,l+\lambda} = \frac{1}{n-\lambda} \sum_{i=1}^{n-\lambda} \tilde{R}_{i,i+\lambda} \quad , \qquad 1 \leqq \lambda \leqq n-1 \quad ,$$

where $\tilde{\boldsymbol{R}} = \| \tilde{R}_{jk} \|$. More explicitly, in terms of the observation vectors $\boldsymbol{z}_k$, $1 \leqq k \leqq m$ [see (4.2) and (4.6)],

$$\tilde{R}(0) = \frac{1}{mn} \sum_{i=1}^{n} \sum_{k=1}^{m} |z_k(t_i)|^2 \tag{4.11}$$

$$\tilde{R}(t_i - t_{i+\lambda}) = \frac{1}{m(n-\lambda)} \sum_{i=1}^{n-\lambda} \sum_{k=1}^{m} z_k(t_i)\bar{z}_k(t_{i+\lambda}) \quad , \qquad 1 \leqq \lambda \leqq n-1 \quad . \tag{4.12}$$

Of course this solution is possible only if

$$p = 2n - 1 \tag{4.13}$$

since (4.11) represents one real equation and (4.12) is equivalent to $2(n-1)$ real equations. We shall exploit this argument in the next section.

Consider now the special case where $\boldsymbol{R}$ is a scalar multiple of the identity matrix (independent of $\boldsymbol{a}$). Then (4.8) reduces to the p scalar equations

$$\operatorname{Re} \frac{\partial \bar{\boldsymbol{c}}'}{\partial \alpha_j} (\tilde{\boldsymbol{Z}} - \boldsymbol{c}) = 0 \quad , \qquad 1 \leqq j \leqq p \quad , \tag{4.14}$$

which now require no conditions on n. If in particular

$$c(t) = \alpha c^*(t)$$

where $c^*(t)$ is not a function of the particular parameter $\alpha \neq 0$, then (4.14) may be solved explicitly for the maximum likelihood estimator $\hat{\alpha}$ of α, viz.:

$$\hat{\alpha} = \frac{1}{|\boldsymbol{c}^*|^2} \operatorname{Re} \bar{\boldsymbol{c}}^{*\prime} \tilde{\boldsymbol{Z}} \quad .$$

[See Section 1, specifically, equation (1.3) .]

If $c(t)$ is a linear function of $\boldsymbol{a}$, say

$$c(t) = \sum_{i=1}^{p} \alpha_i h_i(t) \quad ,$$

then (4.14) becomes

$$\sum_{k=1}^{n} \operatorname{Re} \left[\bar{h}_i(t_k) \tilde{Z}_k - \sum_{j=1}^{p} \alpha_j h_j(t_k) \bar{h}_i(t_k) \right] = 0 \quad , \qquad 1 \leqq i \leqq p \quad ,$$

where

$$\tilde{\boldsymbol{Z}}' = \{\tilde{Z}_1, \cdot \cdot \cdot, \tilde{Z}_n\} \quad .$$

The α_j now may be determined as solutions of the system of linear algebraic equations

$$\sum_{j=1}^{p} \left[\operatorname{Re} \sum_{k=1}^{n} h_j(t_k) \bar{h}_i(t_k) \right] \alpha_j = \operatorname{Re} \sum_{k=1}^{n} \bar{h}_i(t_k) \tilde{Z}_k \quad , \qquad 1 \leqq i \leqq p \quad ,$$

provided the rank of the matrix of the coefficients of the α_j and the rank of the augmented matrix are equal.

Other special cases may be deduced. See, for example, Section 6.

5. Estimation of Random Signals

As our first example let us consider the problem of estimating the parameters in a random signal when there is no deterministic signal present. That is [cf. (4.1)],

$$z(t) = x(t) + y(t) \quad . \tag{5.1}$$

We shall assume that $\{x(t) | -\infty < t < \infty\}$ and $\{y(t) | -\infty < t < \infty\}$, in addition to being independent mean zero complex Gaussian processes, are also stationary,

$$\mathscr{E} x(t+\tau)\bar{x}(\tau) = R_x(t; \boldsymbol{a})$$

and

$$\mathscr{E} y(t+\tau)\bar{y}(\tau) = R_y(t) \quad .$$

For the moment we shall not further specify the parameter $\boldsymbol{a}' = \{\alpha_1, \cdots, \alpha_p\}$ except to assume that $\boldsymbol{a}$ is real and is defined on some p-dimensional region $\mathscr{A}$ of real p-dimensional Euclidean space; and to require that R_x be positive definite for all $\boldsymbol{a} \in \mathscr{A}$ and continuously differentiable with respect to α_j, $1 \leqq j \leqq p$. We shall also assume that R_y is positive definite, and hence

$$R(t; \boldsymbol{a}) = R_x(t; \boldsymbol{a}) + R_y(t) \quad , \tag{5.2}$$

the covariance function of the $\{z(t) \mid -\infty < t < \infty\}$ process, is also positive definite for all $\boldsymbol{a}$ in $\mathscr{A}$.

Let F_x, F_y, F be the spectral distribution functions of the $\{x(t)\}$, $\{y(t)\}$, and $\{z(t)\}$ processes, respectively. We shall assume that

$$\int_{-\infty}^{\infty} f^4 \, dF_x(f) < \infty \tag{5.3}$$

and

$$\int_{-\infty}^{\infty} f^4 \, dF_y(f) < \infty \quad .$$

Thus by Theorem 5.2 of Chapter II

$$\frac{\partial^4}{\partial t^4} R_x(t; \boldsymbol{a})$$

and $R_y^{iv}(t)$ exist (and are continuous) for all t.

Now we shall assume that $\boldsymbol{z}_1, \cdots, \boldsymbol{z}_m$ are m vector observations from a *two*-dimensional Gaussian distribution. Then if t_1 and t_2 are two distinct time points and

$$h = t_1 - t_2 \neq 0 \quad ,$$

the covariance matrix $\boldsymbol{R}_x$ of $\boldsymbol{x}' = \{x(t_1), x(t_2)\}$ is

$$\boldsymbol{R}_x(\boldsymbol{a}) = \begin{bmatrix} R_x(0; \boldsymbol{a}) & R_x(h; \boldsymbol{a}) \\ \bar{R}_x(h; \boldsymbol{a}) & R_x(0; \boldsymbol{a}) \end{bmatrix} ,$$

and the covariance matrix of $\boldsymbol{y}' = \{y(t_1), y(t_2)\}$ is

$$\boldsymbol{R}_y = \begin{bmatrix} R_y(0) & R_y(h) \\ \bar{R}_y(h) & R_y(0) \end{bmatrix} .$$

In brief, then

$$\begin{aligned} c(t) &\equiv 0 \\ n &= 2 \\ m &\quad \text{arbitrary} \\ x(t), y(t) &\quad \text{stationary} \end{aligned}$$

plus the mathematical conditions on the covariance functions stated above.

Let us now specify the parameter $\boldsymbol{a}$. The covariance function R_x of the random signal process is in general complex. We shall write it in polar form as

$$\begin{aligned} R_x(0; \boldsymbol{a}) &= \alpha_3 \\ R_x(h; \boldsymbol{a}) &= \alpha_3 \exp\,(i\phi(\alpha_2 h))\, A(\alpha_1 h) \end{aligned} \tag{5.4}$$

where A is a real even function, $A(0) = 1$, and ϕ is a real odd function, $\phi(0) = 0$. Thus $\boldsymbol{a}$ is a three-dimensional real vector, $p = 3$, and $\boldsymbol{a}' = \{\alpha_1, \alpha_2, \alpha_3\}$. We shall give a physical interpretation of the parameter $\boldsymbol{a}$ a little later. For the present let us compute the estimates and Cramér–Rao bounds for $\boldsymbol{a}$.

Since $n = 2$ and $p = 3$, we have $p = 2n–1$ and the condition of (4.13) is satisfied. Hence from (4.11) and (4.12)

$$\begin{aligned} \tilde{R}(0) &= \frac{1}{2m} \sum_{k=1}^{m} [\,|z_k(t_1)|^2 + |z_k(t_2)|^2] \\ \tilde{R}(h) &= \frac{1}{m} \sum_{k=1}^{m} z_k(t_1)\, \bar{z}_k(t_2) \end{aligned} \tag{5.5}$$

from which follow [see (5.2)] the maximum likelihood estimates

$$A(\hat{\alpha}_1 h) = \frac{|\tilde{R}(h) - R_y(h)|}{\tilde{R}(0) - R_y(0)} \tag{5.6}$$

$$\phi(\hat{\alpha}_2 h) = \arctan \frac{\operatorname{Im}[\tilde{R}(h) - R_y(h)]}{\operatorname{Re}[\tilde{R}(h) - R_y(h)]} \tag{5.7}$$

$$\hat{\alpha}_3 = \tilde{R}(0) - R_y(0) \quad . \tag{5.8}$$

From (4.5)

$$F_{jk} = \operatorname{tr} \boldsymbol{R} \frac{\partial \boldsymbol{R}^{-1}}{\partial \alpha_j} \boldsymbol{R} \frac{\partial \boldsymbol{R}^{-1}}{\partial \alpha_k} \quad , \qquad 1 \leqq j, k \leqq 3 \quad ,$$

where $\boldsymbol{R}(\boldsymbol{a}) = \boldsymbol{R}_x(\boldsymbol{a}) + \boldsymbol{R}_y$, are the elements of the Fisher information matrix. If we write the covariance function $R(h; \boldsymbol{a})$ of the $\{z(t) \mid -\infty < t < \infty\}$ process in polar form as

$$R(h; \boldsymbol{a}) = R_x(h; \boldsymbol{a}) + R_y(h) \equiv B(h) e^{i\psi(h)} \quad ,$$

then some lengthy but elementary calculations yield

$$\begin{aligned} C_{11} = \frac{1}{2h^2 \alpha_3^2 A'^2(\alpha_1 h)} \{ & A^2(\alpha_1 h) [B^2(0) + B^2(h)] + B^2(0) \\ & + B^2(h) \cos 2[\psi(h) - \phi(\alpha_2 h)] \\ & - 4A(\alpha_1 h) B(0) B(h) \cos [\psi(h) - \phi(\alpha_2 h)] \} \end{aligned} \tag{5.9}$$

$$C_{22} = \frac{B^2(0) - B^2(h) \cos 2 [\psi(h) - \phi(\alpha_2 h)]}{2h^2 \alpha_3^2 \phi'^2(\alpha_2 h) A^2(\alpha_1 h)} \tag{5.10}$$

$$C_{33} = \frac{B^2(0) + B^2(h)}{2} \tag{5.11}$$

where $\boldsymbol{F} = \| F_{jk} \|$ is the Fisher information matrix, and we have written

$$\boldsymbol{F}^{-1} = \| C_{jk} \| \quad .$$

Thus

$$\operatorname{Var} \tilde{\alpha}_j \geqq \frac{1}{m} C_{jj} \quad , \qquad 1 \leqq j \leqq 3 \quad , \tag{5.12}$$

for any unbiased estimator $\tilde{\alpha}_j$ of α_j, $1 \leqq j \leqq 3$. In particular, from (5.8) and (5.5)

$$\mathscr{E} \hat{\alpha}_3 = R(0; \boldsymbol{a}) - R_y(0) = R_x(0; \boldsymbol{a}) = \alpha_3 \quad , \tag{5.13}$$

and we see that $\hat{\alpha}_3$ is an unbiased estimator of α_3. But

$$\text{Var } \hat{\alpha}_3 = \frac{B^2(0) + B^2(h)}{2m} \quad .$$

Comparing this with (5.11) and (5.12), we see that $\hat{\alpha}_3$ as given by (5.8) is an efficient statistic for estimating α_3.

Let us attempt to give a physical interpretation of the parameter $\boldsymbol{\alpha}$, and also make some further observations on this model. We shall assume that the spectral distribution function F_x of the $\{x(t) \mid -\infty < t < \infty\}$ process is absolutely continuous, and hence the spectral density W_x equals F_x' a.e. (Lebesgue measure). For simplicity we shall further assume W_x to be continuous. Now define the spectral moments

$$m_k(W_x) = \int_{-\infty}^{\infty} f^k W_x(f)\, df \quad .$$

By (5.3) the moments $m_k(W_x)$ exist for at least $k = 0, 1, 2, 3, 4$. The normalized kth spectral moment of W_x is defined as

$$\mu_k(W_x) = \frac{m_k(W_x)}{m_0(W_x)} \quad , \qquad k = 0, 1, 2 \quad .$$

In particular, $\mu_1(W_x)$ is called the *mean frequency* and

$$\sigma(W_x) = [\mu_2(W_x) - \mu_1^2(W_x)]^{1/2}$$

is called the *spread* or standard deviation.

Now since $R_x(\cdot\, ; \boldsymbol{\alpha})$ and $W_x(\cdot)$ are a Fourier transform pair, (5.4) implies that

$$m_0(W_x) = R_x(0; \boldsymbol{\alpha}) = \alpha_3$$

and we may interpret α_3 as the total power. Furthermore

$$\alpha_2 \phi'(0) = 2\pi \mu_1(W_x)$$

and

$$\alpha_1^2 A^{ii}(0) = -4\pi^2 \sigma^2(W_x) \quad .$$

If we wish to interpret α_2 as mean frequency, we must have

$$\phi'(0) = 2\pi \quad ;$$

and if we wish to interpret α_1 as spread, we must have

$$A^{ii}(0) = -4\pi^2 \quad .$$

Since ϕ has a bounded second derivative in a neighborhood of the origin,

$$\phi(\xi) \approx \phi'(0)\xi$$

and

$$\phi'(\xi) \approx \phi'(0)$$

for $|\xi|$ small; and since A has a bounded third derivative in a neighborhood of the origin,

$$A(\xi) \approx 1 + \tfrac{1}{2}A^{ii}(0)\xi^2$$

and

$$A'(\xi) \approx A^{ii}(0)\xi$$
$$\approx -(\operatorname{sgn} \xi)\ \{-2A^{ii}(0)[1 - A(\xi)]\}^{1/2}$$

for $|\xi|$ small. Hence for $|h|$ small,

$$\phi(\alpha_2 h) \approx 2\pi\alpha_2 h$$
$$\phi'(\alpha_2 h) \approx 2\pi \tag{5.14}$$

and

$$A(\alpha_1 h) \approx 1 - \tfrac{1}{2}(2\pi\alpha_1 h)^2$$
$$A'(\alpha_1 h) \approx -2\pi(\operatorname{sgn} h)\ \{2[1 - A(\alpha_1 h)]\}^{1/2}\ . \tag{5.15}$$

Thus for small $|h|$, the estimators (5.6) and (5.7) for α_1 and α_2, the spread and mean, may be written

$$\hat{\alpha}_1 = \frac{1}{\pi|h|\sqrt{2}}\left[1 - \frac{|\tilde{R}(h) - R_y(h)|}{\tilde{R}(0) - R_y(0)}\right]^{1/2}$$

and

$$\hat{\alpha}_2 = \frac{1}{2\pi h} \arctan \frac{\operatorname{Im}[\tilde{R}(h) - R_y(h)]}{\operatorname{Re}[\tilde{R}(h) - R_y(h)]}$$

respectively; while $\phi'(\alpha_2 h)$ and $A'(\alpha_1 h)$ may be replaced by their approximations (5.14) and (5.15) in the expressions (5.9) and (5.10) for the Cramér bounds.

6. Estimation of Deterministic Signals

We consider the problem of estimating the parameters in a deterministic signal of particular form when there is no random signal present. That is [cf. (4.1)]

$$z(t) = c(t) + y(t) \quad .$$

Explicitly we shall assume that

$$c(t) = \alpha_{p-1} \exp [i(\alpha_0 + \alpha_1 t + \cdot \cdot \cdot + \alpha_{p-2} t^{p-2})] \tag{6.1}$$

where $\boldsymbol{\alpha}' = \{\alpha_0, \alpha_1, \cdot \cdot \cdot, \alpha_{p-1}\}$ is the p-dimensional real parameter we wish to estimate. We shall also assume that $\{y(t) \mid -\infty < t < \infty\}$, in addition to being a mean zero complex Gaussian process, is also stationary.

Let $n = p - 1$, and let t_k, $1 \leqq k \leqq n$, where

$$t_1 < t_2 < \cdot \cdot \cdot < t_n \quad ,$$

be n distinct time points. We shall further assume that the covariance matrix $\boldsymbol{R}$ of $\boldsymbol{z}' = \{z(t_1), \cdot \cdot \cdot, z(t_n)\}$ is diagonal,

$$\boldsymbol{R} = \mathscr{E}(\boldsymbol{z} - \boldsymbol{c})(\bar{\boldsymbol{z}} - \bar{\boldsymbol{c}})' = \sigma^2 \boldsymbol{I} \quad , \qquad \sigma^2 > 0 \quad ,$$

where $\boldsymbol{c}' = \{c(t_1), \cdot \cdot \cdot, c(t_n)\}$ and $\boldsymbol{I}$ is the $n \times n$ identity matrix.

In brief, then,

$$\begin{aligned} c(t) &= \alpha_{p-1} \exp \left(i \sum_{j=0}^{p-2} \alpha_j t^j \right) \\ n &= p - 1 \\ x(t) &\equiv 0 \\ \boldsymbol{R} &= \sigma^2 \boldsymbol{I} \quad . \end{aligned}$$

From (4.8) the likelihood equations are

$$\mathrm{Re} \frac{\partial \bar{\boldsymbol{c}}'}{\partial \alpha_j} \sum_{k=1}^{m} (\boldsymbol{z}_k - \boldsymbol{c}) = 0 \quad , \qquad 0 \leqq j \leqq p - 1 \quad , \tag{6.2}$$

where $\boldsymbol{z}_k' = \{z_k(t_1), \cdot \cdot \cdot, z_k(t_n)\}$, $1 \leqq k \leqq m$, are the m independent vector observations [see (4.2)]. If we introduce the notation

$$\tilde{Z}_j = \frac{1}{m} \sum_{k=1}^{m} z_k(t_j) \quad , \qquad 1 \leqq j \leqq p - 1 \quad , \tag{6.3}$$

then the likelihood equations (6.2) take on the explicit form

$$\begin{aligned} \mathrm{Im} \sum_{k=1}^{p-1} t_k^j \bar{c}(t_k) \tilde{Z}_k &= 0 \quad , \qquad 0 \leqq j \leqq p - 2 \quad , \\ \hat{\alpha}_{p-1}^2 &= \frac{1}{p-1} \mathrm{Re} \sum_{k=1}^{p-1} \bar{c}(t_k) \tilde{Z}_k \quad . \end{aligned} \tag{6.4}$$

By inspection

$$\operatorname{Im} \bar{c}(t_k)\tilde{Z}_k = 0 \quad , \qquad 1 \leqq k \leqq p-1 \quad , \tag{6.5}$$

is a solution of the first $p - 1$ equations of (6.4). These in turn may be further expanded to read

$$\hat{\alpha}_0 + \hat{\alpha}_1 t_k + \cdot \cdot \cdot + \hat{\alpha}_{p-2} t_k^{p-2} = \tilde{H}_k \quad , \qquad 1 \leqq k \leqq p-1 \quad , \tag{6.6}$$

where

$$\tilde{H}_k = \arctan \frac{\operatorname{Im} \tilde{Z}_k}{\operatorname{Re} \tilde{Z}_k} \quad , \qquad 1 \leqq k \leqq p-1 \quad .$$

The determination of $\hat{\alpha}_0, \hat{\alpha}_1, \cdot \cdot \cdot, \hat{\alpha}_{p-2}$ now has been reduced to the problem of solving a system of linear algebraic equations with a Vandermode determinant. Since $\operatorname{Im} \bar{c}(t_k)\tilde{Z}_k = 0$ [see (6.5)], we infer that

$$\operatorname{Re} \bar{c}(t_k)\tilde{Z}_k = \pm \; |\bar{c}(t_k)\tilde{Z}_k| = \pm \; \alpha_{p-1} \, |\tilde{Z}_k| \quad ,$$

and hence from (6.4)

$$\hat{\alpha}_{p-1} = \frac{1}{p-1} \sum_{k=1}^{p-1} |\tilde{Z}_k| \tag{6.7}$$

because $\alpha_{p-1} > 0$.

If $\boldsymbol{F} = \| F_{jk} \|$ is the Fisher information matrix, then (4.5) implies that

$$F_{jk} = F_{kj} = \frac{2}{\sigma^2} \operatorname{Re} \frac{\partial \bar{\boldsymbol{c}}'}{\partial \alpha_j} \frac{\partial \boldsymbol{c}}{\partial \alpha_k} \quad , \qquad 0 \leqq j, k \leqq p-1 \quad ,$$

and for our problem

$$\begin{aligned} F_{jk} &= \frac{2}{\sigma^2} \alpha_{p-1}^2 \sum_{\gamma=1}^{p-1} t_\gamma^{j+k} \quad , && 0 \leqq j, k \leqq p-2 \quad , \\ F_{j,\,p-1} &= 0 \quad , && 0 \leqq j \leqq p-2 \quad , \\ F_{p-1,\,p-1} &= \frac{2}{\sigma^2}(p-1) \quad . \end{aligned} \tag{6.8}$$

If we write $\boldsymbol{F}^{-1} = \| C_{jk} \|$, then

$$\operatorname{Var} \tilde{\alpha}_j \geqq \frac{1}{m} C_{jj} \quad , \qquad 0 \leqq j \leqq p-1 \quad ,$$

for any unbiased estimator $\tilde{\alpha}_j$ of α_j, $0 \leqq j \leqq p-1$. Thus from (6.6), (6.7), and (6.8) we may explicitly compute the estimators $\hat{\alpha}_j, 0 \leqq j \leqq p-1$, and the Cramér–Rao lower bounds.

It is possible to compute the mean and variance of $\hat{\alpha}_{p-1}$ [see (6.7)] in closed form. Returning to (6.3) we see that $\tilde{Z}_j$ is normal $(c(t_j), \sigma^2/m)$. Hence from (3.5) of Chapter III the density function of $r_j = |\tilde{Z}_j|$ is

$$g(r_j) = \frac{2mr_j}{\sigma^2} \exp\{-m[r_j^2 + |c(t_j)|^2]/\sigma^2\}\, I_0\left(\frac{2m|c(t_j)|r_j}{\sigma^2}\right) \quad , \qquad r_j \geqq 0 \quad ,$$

where I_ν is the modified Bessel function of the first kind and order ν. Thus, from (4.1) of Chapter III,

$$\begin{aligned}\mathscr{E} r_j &= \frac{\sigma}{\sqrt{m}} \Gamma\left(\frac{3}{2}\right) \exp\left[\frac{-m|c(t_j)|^2}{\sigma^2}\right] {}_1F_1\left(\frac{3}{2}, 1; \frac{m|c(t_j)|^2}{\sigma^2}\right) \\ &= \frac{\sigma}{2}\left(\frac{\pi}{m}\right)^{1/2} \exp\left[\frac{-m|c(t_j)|^2}{2\sigma^2}\right]\left[\left(1 + \frac{m|c(t_j)|^2}{\sigma^2}\right) I_0\left(\frac{m|c(t_j)|^2}{2\sigma^2}\right)\right. \\ &\qquad \left. + \frac{m|c(t_j)|^2}{\sigma^2} I_1\left(\frac{m|c(t_j)|^2}{2\sigma^2}\right)\right] \quad ,\end{aligned}$$

where ${}_1F_1$ is the confluent hypergeometric function, and from (2.10) or (4.1) of Chapter III,

$$\mathscr{E} r_j^2 = \frac{\sigma^2}{m} + |c(t_j)|^2 \quad .$$

Now from (6.7)

$$\mathscr{E} \hat{\alpha}_{p-1} = \frac{1}{p-1} \sum_{k=1}^{p-1} \mathscr{E} r_k$$

and

$$\operatorname{Var} \hat{\alpha}_{p-1} = \left(\frac{1}{p-1}\right)^2 \sum_{k=1}^{p-1} \operatorname{Var} r_k \quad .$$

As an illustration, let us compute the estimators $\hat{\alpha}_k$ and the Cramér–Rao lower bounds C_{jj} in the case $p = 4$. We have from (6.6) and (6.7) that

$$\hat{\alpha}_0 = \frac{\tilde{H}_1 t_2 t_3(t_3 - t_2) - \tilde{H}_2 t_1 t_3(t_3 - t_1) + \tilde{H}_3 t_1 t_2(t_2 - t_1)}{(t_3 - t_1)(t_3 - t_2)(t_2 - t_1)}$$

$$\hat{\alpha}_1 = -\frac{\tilde{H}_1(t_3^2 - t_2^2) - \tilde{H}_2(t_3^2 - t_1^2) + \tilde{H}_3(t_2^2 - t_1^2)}{(t_3 - t_1)(t_3 - t_2)(t_2 - t_1)}$$

$$\hat{\alpha}_2 = \frac{\tilde{H}_1(t_3 - t_2) - \tilde{H}_2(t_3 - t_1) + \tilde{H}_3(t_2 - t_1)}{(t_3 - t_1)(t_3 - t_2)(t_2 - t_1)}$$

$$\hat{\alpha}_3 = \frac{1}{3}[|\tilde{Z}_1| + |\tilde{Z}_2| + |\tilde{Z}_3|]$$

and from (6.8),

$$C_{00} = \frac{\sigma^2}{2\alpha_3^2} \frac{t_1^2 t_2^2 (t_2 - t_1)^2 + t_1^2 t_3^2 (t_3 - t_1)^2 + t_2^2 t_3^2 (t_3 - t_2)^2}{D}$$

$$C_{11} = \frac{\sigma^2}{2\alpha_3^2} \frac{(t_2^2 - t_1^2)^2 + (t_3^2 - t_1^2)^2 + (t_3^2 - t_2^2)^2}{D}$$

$$C_{22} = \frac{\sigma^2}{2\alpha_3^2} \frac{(t_2 - t_1)^2 + (t_3 - t_1)^2 + (t_3 - t_2)^2}{D}$$

$$C_{33} = \frac{\sigma^2}{6}$$

where

$$D = (t_2 - t_1)(t_3 - t_2)(t_3 - t_1)[(t_2 - t_1)t_3^2 + (t_3 - t_2)t_1^2 + (t_1 - t_3)t_2^2] \quad .$$

We make some further observations on this model. If $p = 1$, and $c(t) = \alpha_0$, and if n is arbitrary, then the problem is trivial. If $p = 2$, and $c(t) = \alpha_1 e^{i\alpha_0}$, and if n is arbitrary, then

$$\tan \hat{\alpha}_0 = \frac{\operatorname{Im} \sum_{k=1}^{n} \tilde{Z}_k}{\operatorname{Re} \sum_{k=1}^{n} \tilde{Z}_k}$$

$$\hat{\alpha}_1 = \frac{1}{n} \left| \sum_{k=1}^{n} \tilde{Z}_k \right| \quad ,$$

and

$$C_{00} = \frac{\sigma^2}{2n\alpha_1^2}$$

$$C_{11} = \frac{\sigma^2}{2n} \quad .$$

However, for $p > 2$ and $n > p - 1$, it does not appear possible to solve the likelihood equations in closed form. If $p > 2$ and $n < p - 1$, some of the Cramér bounds become infinite. For example, if $p = 4$, and $c(t) = \alpha_3 \times \exp[i(\alpha_0 + \alpha_1 t + \alpha_2 t^2)]$, and $t_k = (k - 1)h$, $1 \leqq k \leqq n$, then

$$C_{11} = \frac{6\sigma^2}{\alpha_3^2 h^2} \frac{(2n - 1)(8n - 11)}{n(n^2 - 1)(n^2 - 4)}$$

—which becomes infinite if $n < 3$.

VI

SPECTRAL MOMENT ESTIMATION

0. Introduction

The problem of estimating the spectral density of a wide-sense stationary process has received wide attention (see, for example, [An], [BT], [GR], [H].) Many practical problems, however, really require a knowledge only of certain functionals of the spectrum. In other cases one has no alternative but to estimate particular parameters of the spectral density function (as some measure of the character of the population spectrum) rather than the spectral density itself. Of particular interest in various physical applications are the spectral moments of a stochastic process. In this connection we cite (in addition to some of the references quoted in the Preface) [BG], [BRS], [EE], [La], [Le], [N], [Ra], [Ri], [S], [SBS].

We recall that the definition of spectral moment was introduced in Chapter V (see page 153). Thus if $\{x(t) \mid -\infty < t < \infty\}$ is a wide-sense stationary process with integrable spectral density function W, we define the kth spectral moment (assumed to exist) to be

$$m_k(W) = \int_{-\infty}^{\infty} f^k W(f)\,df \quad . \tag{0.1}$$

The normalized kth spectral moment of W is defined as

$$\mu_k(W) = \frac{m_k(W)}{m_0(W)} \quad . \tag{0.2}$$

In particular, $\mu_1(W)$ is called the mean frequency and

$$\sigma(W) = [\mu_2(W) - \mu_1^2(W)]^{1/2} \tag{0.3}$$

is called the r.m.s. spread or spectral width.

We devote this chapter to a discussion of the problem of choosing suitable estimators of the spectral moments μ_1 and σ. There are numerous ways of attacking this problem. In Section 1 we investigate the basic spectral and covariance approaches, dwelling more on motivation and discussion than on mathematical rigor. In Sections 2 and 3 we survey some other possibilities. Later sections will concentrate on a particular technique, namely, the coherent spectral approach. In particular, we shall explicitly compute the mean and variance of the estimators of μ_1 and σ, paying appropriate attention to the technical details.

1. Spectral and Covariance Approaches

Let $\{x(t) \mid -\infty < t < \infty\}$ be a mean zero wide-sense stationary complex stochastic process with continuous covariance function R and absolutely continuous spectral distribution function S. Then the spectral density function $W = S'$ a.e. (Lebesgue measure). We shall assume further that

$$\int_{-\infty}^{\infty} f^4 W(f) df < \infty \quad .$$

Thus the spectral moments $m_k(W)$ exist for at least $k = 0, 1, 2, 3, 4$. We also find it convenient to define $\mu_k(q)$ as

$$\mu_k(q) = \frac{\int_{-\infty}^{\infty} t^k q(t) dt}{\int_{-\infty}^{\infty} q(t) dt} \tag{1.1}$$

for any nonnegative integrable function q with positive area whose kth moment exists. This notation is consistent with our definition of $\mu_k(W)$, see (0.1) and (0.2). Furthermore, if q is bounded and absolutely integrable on $(-\infty, \infty)$, then

$$\int_{-\infty}^{\infty} |q(t)|^2 dt = \int_{-\infty}^{\infty} |Q(f)|^2 df < \infty$$

where Q is the Fourier transform of q.

Now it seems natural to base estimators of $\mu_k(W)$ on estimators of W itself—or, because the spectral density and covariance function R are a Fourier transform pair—on estimates of R. Since the spectral moments $m_k(W)$ are simply related to the derivatives of $R(h)$ at h equal to zero, the moments may be evaluated from the covariance and cross-covariance functions of the random process $\{x(t)\}$ with its derivative processes $\{x^{(k)}(t)\}$. We shall discuss the spectral and covariance approaches to spectral moment estimation in the present section. In Sections 2 and 3 we shall consider other assaults on the estimation problem.

The most direct approach to the problem is to base estimators of $\mu_1(W)$ on estimates of W itself. For example, since

$$\mu_1(W) = \frac{\int_{-\infty}^{\infty} f W(f) df}{\int_{-\infty}^{\infty} W(f) df} \quad , \tag{1.2}$$

it seems reasonable to choose an estimate of W, say $\tilde{W}$, and then use $\mu_1(\tilde{W})$ as an estimate of $\mu_1(W)$. Now in the literature, a frequently used estimator of $W(f)$ is

$$\tilde{W}_n(f) = \frac{1}{2T} \left| \int_{-T}^{T} x_n(t) e^{-i2\pi f t} dt \right|^2 \tag{1.3}$$

where x_n is an observation of length $2T$ on $\{x(t) \mid -\infty < t < \infty\}$. More generally we may write (1.3) as

$$\zeta_n(f) = \left| \int_{-\infty}^{\infty} p(t) x_n(t) e^{-i2\pi f t} dt \right|^2 \tag{1.4}$$

where p is a weighting function. [We shall assume that p is a real, piecewise continuous, nonnegative, bounded, and absolutely integrable function on $(-\infty, \infty)$.] If

$$p(t) = \frac{1}{\sqrt{2T}} \quad , \qquad |t| < T \quad ,$$

$$p(t) = 0 \quad , \qquad |t| > T \quad ,$$

then (1.4) reduces to (1.3). Certainly, then, if x_n, $1 \leqq n \leqq N$, are N independent observations on $\{x(t) \mid -\infty < t < \infty\}$,

$$c(f) = \frac{1}{N} \sum_{n=1}^{N} \zeta_n(f) \tag{1.5}$$

is a candidate for an estimator of $W(f)$.

From (1.4)

$$\mathscr{E} c(f) = \int_{-\infty}^{\infty} W(g) |P(f-g)|^2 dg \tag{1.6}$$

where P is the Fourier transform of p. The relation of the form of (1.6) to the (unnormalized) density function of the sum of two independent random variables immediately leads to the conclusion that

$$\mu_1(\mathscr{E} c) = \mu_1(W) + \mu_1(|P|^2) \quad . \tag{1.7}$$

For p real, $\mu_1(|P|^2) = 0$, and hence $\mu_1(c)$ is a reasonable estimator of $\mu_1(W)$. We shall return to this statistic later.

Now we recall that the spectral density function W and the covariance function R are a Fourier transform pair. Thus from

$$\mathscr{E}x(t)\bar{x}(s) = R(t-s) = \int_{-\infty}^{\infty} W(f)e^{i2\pi(t-s)f}df \quad , \tag{1.8}$$

we may differentiate to obtain

$$\mathscr{E}x'(t)\bar{x}(s) = R'(t-s) = i2\pi \int_{-\infty}^{\infty} fW(f)e^{i2\pi(t-s)f}df \quad . \tag{1.9}$$

Then if we define

$$\eta_n(f) = -i\int_{-\infty}^{\infty} p(t)x_n'(t)e^{-i2\pi ft}dt \; \overline{\int_{-\infty}^{\infty} p(s)x_n(s)e^{-i2\pi fs}ds}$$

we have

$$\mathscr{E}\eta_n(f) = 2\pi\int_{-\infty}^{\infty} gW(g)\,|P(f-g)|^2 dg \quad .$$

These last two equations, of course, imply that the derivative process $\{x'(t) \mid -\infty < t < \infty\}$ is given. Now let

$$b(f) = \frac{1}{2\pi N}\sum_{n=1}^{N}\eta_n(f) \quad .$$

Then

$$\mu_1(W) = \frac{\int_{-\infty}^{\infty} \mathscr{E}b(f)df}{\int_{-\infty}^{\infty} \mathscr{E}c(f)df} \quad .$$

Hence

$$\frac{\int_{-\infty}^{\infty} b(f)df}{\int_{-\infty}^{\infty} c(f)df}$$

is another candidate for an estimator of $\mu_1(W)$.

Our discussion of estimators, so far, has been concentrated on spectral approaches. Now

$$\tilde{R}_n(h) = \frac{1}{2T}\int_{-T}^{T} x_n(t+h)\bar{x}_n(t)dt \tag{1.10}$$

is an unbiased estimate of $R(h)$. Let us examine this statistic with a view to using it (or some modification of it) as an estimator of $\mu_1(W)$. First we see that (1.10) does not restrict the value of h and $\tilde{R}_n(h)$ fails to equal $\bar{\tilde{R}}(-h)$—an undesirable state of affairs. To overcome these difficulties we could use

$$R_n^*(h) = \frac{1}{2T - |h|} \int_{-T+\frac{1}{2}|h|-\frac{1}{2}h}^{T-\frac{1}{2}|h|-\frac{1}{2}h} x_n(t+h)\bar{x}_n(t)\,dt \tag{1.11}$$

in place of $\tilde{R}_n(h)$. Then $R_n^*(h) = \overline{R_n^*}(-h)$; the argument of $x_n(\cdot)$ remains in the interval $[-T, T]$; the argument of $R_n^*(\cdot)$ remains in the interval $(-2T, 2T)$. Of course, if $|h|$ is small compared with T (as is frequently the case), we may use $\tilde{R}_n(h)$ in place of $R_n^*(h)$. (It may be shown that such an assumption greatly simplifies the calculations.) More generally we shall consider

$$\psi_n(h) = \frac{1}{\Pi(h)} \int_{-\infty}^{\infty} p(t+h)p(t)x_n(t+h)\bar{x}_n(t)\,dt \tag{1.12}$$

where

$$\Pi(h) = \int_{-\infty}^{\infty} p(t+h)p(t)\,dt \tag{1.13}$$

and p is a real weighting function. Then $\psi_n(h)$ is an unbiased estimator of $R(h)$; $\psi_n(h) = \bar{\psi}_n(-h)$; and if

$$p(t) = 1 \quad , \qquad |t| < T \quad ,$$
$$p(t) = 0 \quad , \qquad |t| > T \quad ,$$

then $\psi_n(h) = R_n^*(h)$.

From (1.9) we conclude that

$$\mu_1(W) = \frac{1}{i2\pi}\frac{R'(0)}{R(0)} = \frac{1}{i2\pi}\frac{\mathscr{E}x'(t)\bar{x}(t)}{\mathscr{E}|x(t)|^2} \quad .$$

Now if the $\{x'(t) \mid -\infty < t < \infty\}$ process is given,

$$\chi_n(h) = \frac{1}{i2\pi\Pi(h)} \int_{-\infty}^{\infty} p(t+h)p(t)x_n'(t+h)\bar{x}_n(t)\,dt$$

is an unbiased estimator of

$$\int_{-\infty}^{\infty} fW(f)e^{i2\pi hf}\,df \quad .$$

Hence

$$\frac{\frac{1}{N}\sum_{n=1}^{N}\chi_n(0)}{\frac{1}{N}\sum_{n=1}^{N}\psi_n(0)}$$

is an estimator of $\mu_1(W)$ based on a covariance approach.

We also remark that if we write the covariance function R of the $\{x(t) \mid -\infty < t < \infty\}$ process in polar form as

$$R(h) = A(h)e^{i2\pi\phi(h)} \quad , \tag{1.14}$$

then

$$\mu_1(W) = \phi'(0) \quad .$$

Hence we may base an estimator of $\mu_1(W)$ on an estimate of the first derivative of the phase of the covariance function evaluated at $h = 0$. If the derivative process $\{x'(t)\}$ is not available, we may write, for $|h|$ small and nonzero,

$$\mu_1(W) = \phi'(0) \approx \frac{\phi(h) - \phi(0)}{h} = \frac{\phi(h)}{h}$$
$$= \frac{1}{2\pi h} \arctan \frac{\text{Im } R(h)}{\text{Re } R(h)} \quad .$$

Of the various candidates for estimating $\mu_1(W)$ that have been advanced, the only one that does not involve derivative processes is the c statistic based on the ζ_n[see (1.4) and (1.5)]. Since the derivative process $\{x'(t) \mid -\infty < t < \infty\}$ is generally not available, we shall choose ζ_n as our basic statistic. Furthermore, since ζ_n is the Fourier transform of $\Pi\psi_n$, we shall also admit ψ_n as a prominent statistic. An analysis based on

$$\hat{\mu}_1(W) \equiv \mu_1(c) \tag{1.15}$$

as an estimator of $\mu_1(W)$ is called a *spectral approach;* an analysis based on

$$\hat{M}_1(W) = \frac{1}{2\pi h} \arctan \frac{\text{Im } \hat{R}(h)}{\text{Re } \hat{R}(h)} \quad , \tag{1.16}$$

where

$$\hat{R}(h) = \frac{1}{N} \sum_{n=1}^{N} \psi_n(h) \tag{1.17}$$

is an unbiased estimator of $R(h)$, is called a *covariance approach.*

Our discussion seems to indicate that (1.4) and (1.12) are the most general statistics. However, one may attempt to construct an estimate of W from an estimate of R (by taking transforms) and conversely. In fact, more generally, we may consider the Fourier transform of a *weighted* estimator of $\psi_n(t)$, say

$$\omega_n(f) = \int_{-\infty}^{\infty} q(t)\psi_n(t)e^{-i2\pi ft}dt \quad , \tag{1.18}$$

where q is a real weighting function. Since ω_n is a candidate for an estimator of W, and since W is real, it is desirable to have ω_n real. This will be true if q is an even function. We shall make this assumption. (Note that ζ_n is a special case of ω_n when $q = \Pi$.) It might be conjectured that ω_n is a better estimator than ζ_n. This is not true. To prove it, let

$$z(f) = \frac{1}{N}\sum_{n=1}^{N} \omega_n(f) \quad .$$

Then the inverse transform of ω_n [see (1.18)] leads to

$$q(0)\psi_n'(0) = i2\pi \int_{-\infty}^{\infty} f\omega_n(f)df$$

and hence

$$\mu_1(z) = \frac{\frac{1}{N}\sum_{n=1}^{N}\int_{-\infty}^{\infty} f\omega_n(f)df}{\frac{1}{N}\sum_{n=1}^{N}\int_{-\infty}^{\infty} \omega_n(f)df} = \frac{1}{i2\pi}\,\frac{\sum_{n=1}^{N}\psi_n'(0)}{\sum_{n=1}^{N}\psi_n(0)} \quad .$$

But from (1.12), the Fourier transform of $\Pi\psi_n$ is ζ_n. Hence

$$\Pi(h)\psi_n(h) = \int_{-\infty}^{\infty} \zeta_n(f)e^{i2\pi hf}df \quad ,$$

and [see (1.5)]

$$\mu_1(c) = \frac{\frac{1}{N}\sum_{n=1}^{N}\int_{-\infty}^{\infty} f\zeta_n(f)df}{\frac{1}{N}\sum_{n=1}^{N}\int_{-\infty}^{\infty} \zeta_n(f)df} = \frac{1}{i2\pi}\,\frac{\sum_{n=1}^{N}\psi_n'(0)}{\sum_{n=1}^{N}\psi_n(0)}$$

—which is identical with $\mu_1(z)$. In other words, the spectral moment estimate of $\mu_1(W)$ is independent of q, and we might just as well let $q = \Pi$ and use ζ_n. Nothing new is gained by using ω_n in place of ζ_n.

While most of the comments in this section were directed toward estimates of $\mu_1(W)$, they apply equally well to estimates of $\sigma(W)$. Thus from (1.6), and for the same reason, we conclude that

$$\sigma^2(\mathscr{E}c) = \sigma^2(W) + \sigma^2(|P|^2) \quad .$$

However, for this case, $\sigma^2(|P|^2) \neq 0$. Thus

$$\sigma(W) = [\sigma^2(\mathscr{E}c) - \sigma^2(|P|^2)]^{1/2}$$

and we may choose

$$\hat{\sigma}(W) = [\sigma^2(c) - \sigma^2(|P|^2)]^{1/2} \tag{1.19}$$

as a reasonable spectral estimator of $\sigma(W)$.

We also note [see (1.18) and (1.12)] that

$$\sigma^2(\mathscr{E}\omega_n) = \sigma^2(W) + \sigma^2(Q)$$

where Q is the Fourier transform of q. Then using the same type of argument as we did to prove that $\mu_1(z) = \mu_1(c)$, we may show that

$$[\sigma^2(z) - \sigma^2(Q)]^{1/2}$$

is identical with $\hat{\sigma}(W)$. Thus, again, no generality is gained by using ω_n in place of ζ_n.

To deduce a reasonable covariance estimator of $\sigma(W)$ we recall that, in general,

$$\mu_k(W) = \frac{1}{(i2\pi)^k} \frac{R^{(k)}(0)}{R(0)}$$

and hence from (1.14)

$$\sigma^2(W) = \mu_2(W) - \mu_1^2(W) = -\frac{1}{4\pi^2} \frac{A^{ii}(0)}{A(0)}$$

(since A is an even function and ϕ is odd.) But for $|h|$ small and nonzero,

$$A^{ii}(0) \approx \frac{A(h) - 2A(0) + A(-h)}{h^2} = -\frac{2[A(0) - A(h)]}{h^2}$$

and therefore

$$\sigma^2(W) \approx \frac{1}{2\pi^2 h^2}\left[1 - \frac{A(h)}{A(0)}\right].$$

Thus we may choose

$$\hat{\Sigma}(W) = \frac{1}{\pi |h| \sqrt{2}}\left[1 - \frac{|\hat{R}(h)|}{\hat{R}(0)}\right]^{1/2}, \tag{1.20}$$

where $\hat{R}(h)$ is an unbiased estimator of $R(h)$ [see (1.17)], as a reasonable covariance estimator of $\sigma(W)$.

2. The Addition of Noise

In the previous section we considered the basic approaches to the problem of estimating the spectral moments of a mean zero wide-sense stationary complex stochastic process $\mathscr{X} = \{x(t) \mid -\infty < t < \infty\}$. We now shall consider certain variations on the problem. One such ramification concerns the addition of noise. For example, in the analysis of the velocity distribution of turbulent phenomena, it becomes necessary to estimate the spectral moments of $\mathscr{X}$ when $\mathscr{X}$ is immersed in additive wide-sense stationary complex noise, say $\mathscr{N} = \{v(t) \mid -\infty < t < \infty\}$. We shall assume that $\mathscr{N}$ has mean zero and continuous covariance function R_v, and is independent of $\mathscr{X}$. Then our (independent) observations will be of the form

$$y_n(t) = x_n(t) + v_n(t) \quad , \qquad 1 \leqq n \leqq N \quad , \tag{2.1}$$

where $\mathscr{Y} = \{y(t) \mid -\infty < t < \infty\} = \mathscr{X} + \mathscr{N}$. Our problem is still the same, namely, to estimate $\mu_1(W)$ and $\sigma(W)$ where W is the spectral density of $\mathscr{X}$. But now our analysis must be based on the $y_n(t)$, $1 \leqq n \leqq N$, samples. We shall show that such a modification of the problem does not fundamentally change the theory.

Let us first analyze this new problem by the spectral approach. Toward this end we define

$$\begin{aligned} \zeta_n^*(f) &= \left| \int_{-\infty}^{\infty} p(t) y_n(t) e^{-i2\pi ft} dt \right|^2 \\ &= \int_{-\infty}^{\infty} \Pi(t) \psi_n^*(t) e^{-i2\pi ft} dt \end{aligned} \tag{2.2}$$

where

$$\psi_n^*(t) = \frac{1}{\Pi(t)} \int_{-\infty}^{\infty} p(s+t) p(s) y_n(s+t) \bar{y}_n(s) ds \tag{2.3}$$

and p and Π are as defined in Section 1. Similarly, if we let

$$\gamma(f) = \left| \int_{-\infty}^{\infty} p(t) v(t) e^{-i2\pi ft} dt \right|^2 \quad ,$$

then we may define c^* as

$$c^*(f) = \frac{1}{N} \sum_{n=1}^{N} \zeta_n^*(f) - \mathscr{E}\gamma(f)$$

and

$$\mathscr{E}c^*(f) = \mathscr{E}c(f) = \int_{-\infty}^{\infty} W(g) |P(f-g)|^2 dg$$

[see (1.6)] where c and ζ_n are defined as in (1.5) and (1.4). Suitable estimators (on a spectral basis) of $\mu_1(W)$ and $\sigma(W)$ are therefore

$$\hat{\mu}_1(W) = \mu_1(c^*)$$

and

$$\hat{\sigma}(W) = [\sigma^2(c^*) - \sigma^2(|P|^2)]^{1/2} \quad ,$$

respectively.

The introduction of the weighting function q [see (1.18)] again yields nothing new. For, if we define z^* as

$$z^*(f) = \frac{1}{N} \sum_{n=1}^{N} \omega_n^*(f) - \mathscr{E}\Omega(f)$$

where

$$\omega_n^*(f) = \int_{-\infty}^{\infty} q(t)\psi_n^*(t)e^{-i2\pi ft}dt$$

and

$$\Omega(f) = \int_{-\infty}^{\infty} q(t)\left[\frac{1}{\Pi(t)}\int_{-\infty}^{\infty} p(s+t)p(s)v(s+t)\bar{v}(s)ds\right]e^{-i2\pi ft}dt \quad ,$$

then, as in the previous section, we may show that

$$\hat{\mu}_1(W) = \mu_1(z^*)$$

and

$$\hat{\sigma}(W) = [\sigma^2(z^*) - \sigma^2(Q)]^{1/2} \quad .$$

Thus we may return to the ζ_n^*.

Frequently, as in the case of radar measurements of ionized wakes, observations on the noise process $\mathscr{N}$ are available before and after the appearance of the wake. In such cases we may replace c^* by

$$c^{**}(f) = \frac{1}{N} \sum_{n=1}^{N} \zeta_n^*(f) - \frac{1}{M} \sum_{m=1}^{M} \gamma_m(f)$$

where γ_m, $1 \leqq m \leqq M$, are M samples on γ independent of the ζ_n^*, $1 \leqq n \leqq N$, observations. Presumably M will be much larger than N. Then, of course, the estimators of $\mu_1(W)$ and $\sigma(W)$ (on a spectral basis) become

$$\mu_1(c^{**})$$

and

$$[\sigma^2(c^{**}) - \sigma^2(|P|^2)]^{1/2} \quad ,$$

respectively.

We now turn to a consideration of a covariance approach to the problem of estimating spectral moments with noise. Let

$$\hat{R}_y(h) = \frac{1}{N} \sum_{n=1}^{N} \psi_n^*(h)$$

where ψ_n^* is as defined by (2.3). Then $\hat{R}_y$ is an unbiased estimator of R_y, the covariance function of the $\mathscr{Y}$ process. Hence

$$\hat{M}_1(W) = \frac{1}{2\pi h} \arctan \frac{\text{Im}\,[\hat{R}_y(h) - R_v(h)]}{\text{Re}\,[\hat{R}_y(h) - R_v(h)]}$$

and

$$\hat{\Sigma}(W) = \frac{1}{\pi\,|h|\,\sqrt{2}} \left[1 - \frac{|\hat{R}_y(h) - R_v(h)|}{\hat{R}_y(0) - R_v(0)}\right]^{1/2}$$

are suitable estimators of $\mu_1(W)$ and $\sigma(W)$, respectively (on a covariance basis).

3. Noncoherent Statistics

The analyses of the previous two sections were concerned with *coherent* estimation, that is, we utilized the amplitudes *and* phases of the observations x_n(or y_n) on the stochastic process. Sometimes, however, one cannot measure x_n, but perhaps only $|x_n|^2$. And since $X_n = |x_n|^2$ is a magnitude (amplitude squared, but no phase), techniques based on the X_n samples are generally referred to as *noncoherent* estimation methods.

Suppose now we define

$$X(t) = |x(t)|^2$$

where, as in earlier sections, $\{x(t)\,|\,-\infty < t < \infty\}$ is a mean zero wide-sense stationary complex stochastic process with covariance function R and spectral density W. Furthermore, we shall make the crucial assumption that the $\{x(t)\,|\,-\infty < t < \infty\}$ process is Gaussian. We wish to estimate the spectral moments of W based on observations on $\{X(t)\,|\,-\infty < t < \infty\}$. Clearly one cannot estimate $\mu_1(W)$, since all phase information has been eliminated. However, we can estimate $\sigma(W)$; because if W_X is the spectral density of the $\{X(t)\,|\,-\infty < t < \infty\}$ process, then

$$\sigma^2(W_X) = \mu_2(W_X) = 2\sigma^2(W) \quad . \tag{3.1}$$

We also observe that if R_X is the covariance function of the $\{X(t)\,|\,-\infty < t < \infty\}$ process, then

$$R_X(h) = |R(h)|^2 \quad . \tag{3.2}$$

Now define c_X for the $\{X(t) | -\infty < t < \infty\}$ process analogously to the way c was defined for the $\{x(t) | -\infty < t < \infty\}$ process [see (1.5), (1.4), (2.2), and (2.3)]. That is, we replace x_n by $X_n^* = X_n - \mathscr{E}X$ (so that X_n^* has mean zero) and let

$$\begin{aligned} c_X(f) &= \frac{1}{N} \sum_{n=1}^{N} \left| \int_{-\infty}^{\infty} p(t) X_n^*(t) e^{-i2\pi ft} dt \right|^2 \\ &= \frac{1}{N} \sum_{n=1}^{N} \int_{-\infty}^{\infty} \Pi(t) \Psi_n(t) e^{-i2\pi ft} dt \end{aligned}$$

where

$$\Psi_n(t) = \frac{1}{\Pi(t)} \int_{-\infty}^{\infty} p(s+t) p(s) X_n^*(s+t) X_n^*(s) ds$$

and Π has been defined earlier [see (1.13)]. Then, for example,

$$\mathscr{E} c_X(f) = \int_{-\infty}^{\infty} W_X(g) | P(f-g) |^2 dg$$

and (3.1) allows us to estimate $\sigma(W)$ by the formula

$$\hat{\sigma}_X(W) = \{\tfrac{1}{2}[\sigma^2(c_X) - \sigma^2(|P|^2)]\}^{1/2} \quad .$$

That is, $\hat{\sigma}_X(W)$ is an estimate of $\sigma(W)$ based on a noncoherent spectral approach.

From a covariance viewpoint

$$\hat{\Sigma}_X(W) = \frac{1}{2\pi |h|} \left[1 - \frac{\hat{R}_X(h)}{\hat{R}_X(0)}\right]^{1/2}$$

is an estimator of $\sigma(W)$ where

$$\hat{R}_X(h) = \frac{1}{N} \sum_{n=1}^{N} \Psi_n(h)$$

is an unbiased estimate of $R_X(h)$.

There appears to be no simple choice of statistic in the noncoherent case when noise is added. The reason is due to the nonadditivity of the covariance functions of the noncoherent processes. More explicitly, let

$$y(t) = x(t) + \nu(t)$$

where $\{\nu(t) | -\infty < t < \infty\}$ is mean zero stationary complex Gaussian noise independent of x. Let R_N be the covariance function of the $\{|\nu(t)|^2 | -\infty <$

$t < \infty\}$ noise process, and R_Y the covariance function of the $\{|y(t)|^2| -\infty < t < \infty\}$ sum process. Then, in general,

$$R_Y(h) \neq R_X(h) + R_N(h) \quad .$$

In fact,

$$R_Y(h) = R_X(h) + 2\,\mathrm{Re}[R(h)\bar{R}_\nu(h)] + R_N(h) \tag{3.3}$$

where R_ν is the covariance function of the noise process $\{\nu(t)\,|-\infty < t < \infty\}$. If certain additional assumptions are made, for example, if $R(h)$ and $R_\nu(h)$ are assumed to have the same phase, then progress may be made. For example, in such a case (3.3) becomes

$$\begin{aligned} R_Y(h) &= |R(h) + R_\nu(h)|^2 \\ &= [|R(h)| + |R_\nu(h)|]^2 \\ &= [R_X^{1/2}(h) + R_N^{1/2}(h)]^2 \end{aligned}$$

or

$$R_Y^{1/2}(h) = R_X^{1/2}(h) + R_N^{1/2}(h) \quad .$$

Then

$$\tilde{\Sigma}(W) = \frac{1}{\pi|h|\sqrt{2}}\left[1 - \frac{\hat{R}_Y^{1/2}(h) - R_N^{1/2}(h)}{\hat{R}_Y^{1/2}(0) - R_N^{1/2}(0)}\right]^{1/2}$$

and

$$\tilde{\tilde{\Sigma}}(W) = \frac{1}{2\pi|h|}\left[1 - \frac{[\hat{R}_Y^{1/2}(h) - R_N^{1/2}(h)]^2}{[\hat{R}_Y^{1/2}(0) - R_N^{1/2}(0)]^2}\right]^{1/2}$$

are potential candidates for estimators of $\sigma(W)$ (on a covariance basis) where $\hat{R}_Y(h)$ is an unbiased estimator of $R_Y(h)$.

The cases considered above and in the previous two sections do not exhaust the possibilities. For example, it is also possible to deduce estimators of $\mu_1(W)$ and $\sigma(W)$ using the *phase* alone of a complex process, rather than the amplitude or amplitude *and* phase as we have discussed earlier. Also, one could consider *correlated* rather than independent observations. For example, a record of data of length L might be decomposed into N samples x_n, $1 \leqq n \leqq N$, each of length $2T = L/N$. Then certainly it is highly unlikely that the x_n would be independent or even uncorrelated. Another practical model would be to treat discrete parameter stochastic processes (time series) rather than the continuous parameter types we have dealt with exclusively in this chapter. And, of course, one could apply the results to real instead of complex processes. Naturally, in this case there would be no $\mu_1(W)$ to estimate, since the spectral density is an even function; and the distinction between coherent and noncoherent becomes mainly academic.

In fact, almost every combination of spectral or covariance approaches, no noise or additive noise, coherent or noncoherent or phase techniques, independent or correlated samples, continuous or discrete parameter systems, complex or real processes, could be used. For example, the spectral, noisy, coherent, independent, continuous, complex case has been treated in [MR 1], while the covariance, noisy, coherent, independent, continuous, complex case has been treated in [MR 2]. Also, the spectral, no noise, correlated, discrete, real case has been treated in [MR 3].

4. Statistics of the Spectral Moment Estimators

What we desire to do in this section is to compute the statistics of the estimators of the mean frequency and spectral width when these estimators are based on a spectral approach. Specifically we wish to compute the mean and variance of

$$\hat{\mu}_1(W) = \mu_1(c) \tag{4.1}$$

[see (1.15)], an estimator of the mean frequency $\mu_1(W)$, and

$$\hat{\sigma}(W) = [\sigma^2(c) - \sigma^2(|P|^2)]^{1/2} \tag{4.2}$$

[see (1.19)], an estimator of the spectral width $\sigma(W)$. As in Section 1 we start with a mean zero wide-sense stationary complex stochastic process $\{x(t) | -\infty < t < \infty\}$ with covariance function R and spectral density W. We shall also make the assumption that the $\{x(t) | -\infty < t < \infty\}$ process is Gaussian. Then if x_n, $1 \leqq n \leqq N$, are N independent observations on $\{x(t) | -\infty < t < \infty\}$, and if p is a real weighting function, we define

$$\zeta_n(f) = \left| \int_{-\infty}^{\infty} p(t) x_n(t) e^{-i2\pi ft} dt \right|^2$$

[see (1.4)] and

$$c(f) = \frac{1}{N} \sum_{n=1}^{N} \zeta_n(f) \tag{4.3}$$

[see (1.5)]. Then if we let P be the Fourier transform of p, all the symbols appearing in (4.1) and (4.2) will be defined [see (0.1), (0.2), (0.3), and (1.1)].

We shall formally derive formulas for the mean and variance of $\hat{\mu}_1(W)$ and $\hat{\sigma}(W)$ which are asymptotic with the sample size N. The same arguments used in this section may also be employed to derive the corresponding statistics for the various modifications (for example, the covariance approach) of the problem discussed in earlier sections. In Section 5 we shall give a mathematical justification.

To carry out this program let

$$\mathscr{E}c(f) = \gamma(f, f) \quad . \tag{4.4}$$

Then

$$\operatorname{Cov}(c(f), c(g)) = \frac{1}{N} |\gamma(f, g)|^2 \tag{4.5}$$

where

$$\gamma(f, g) = \int_{-\infty}^{\infty} P(f-h)P(h-g)W(h)dh \tag{4.6}$$

[and hence $\gamma(f, f) = \int_{-\infty}^{\infty} |P(f-h)|^2 W(h)dh$, see (1.6)]. Now if we define

$$\xi(f) = c(f) - \gamma(f, f) \quad ,$$

then ξ has mean zero. We also let

$$\begin{aligned} B_k &= \int_{-\infty}^{\infty} f^k \mathscr{E}c(f)df \\ &= \int_{-\infty}^{\infty} f^k \gamma(f, f)df \\ &= B_0 \mu_k(\gamma) \end{aligned} \tag{4.7}$$

and

$$Z_k = \int_{-\infty}^{\infty} f^k \xi(f)df \quad . \tag{4.8}$$

Then Z_k also has mean zero, and by definition [see (4.1)]

$$\hat{\mu}_1(W) = \mu_1(c) = \frac{\int_{-\infty}^{\infty} fc(f)df}{\int_{-\infty}^{\infty} c(f)df} = \frac{Z_1 + B_1}{Z_0 + B_0}$$

$$= \frac{B_1}{B_0} + \frac{1}{B_0^2}(B_0 Z_1 - B_1 Z_0) + \frac{1}{B_0^3}(B_1 Z_0^2 - B_0 Z_0 Z_1) + \cdots \quad . \tag{4.9}$$

It therefore appears that the statistics of $Z_j Z_k$ are of paramount importance. Thus we introduce the notation

$$T_{j,k} = N\mathscr{E} Z_j Z_k = N\mathscr{E} \int_{-\infty}^{\infty} \int_{-\infty}^{\infty} f^j g^k \xi(f)\xi(g)dfdg$$

$$= N \int_{-\infty}^{\infty} \int_{-\infty}^{\infty} f^j g^k \operatorname{Cov}(c(f), c(g)) df dg \tag{4.10}$$
$$= \int_{-\infty}^{\infty} \int_{-\infty}^{\infty} f^j g^k |\gamma(f, g)|^2 df dg$$

[see (4.5).] Furthermore, since

$$B_0 = \int_{-\infty}^{\infty} |P(f)|^2 df \int_{-\infty}^{\infty} W(g) dg$$
$$B_1 = B_0 \mu_1(W) \tag{4.11}$$
$$B_2 = B_0[\mu_2(W) + \mu_2(|P|^2)]$$

[see (4.7)], we find from (4.9) that

$$\mathscr{E}\hat{\mu}_1(W) = \mu_1(W) - \frac{1}{NB_0^2}[T_{0,1} - \mu_1(W)T_{0,0}] + 0(N^{-2}) \tag{4.12}$$

and

$$\operatorname{Var} \hat{\mu}_1(W) = \frac{1}{NB_0^2}[T_{1,1} - 2\mu_1(W)T_{0,1} + \mu_1^2(W)T_{0,0}] + 0(N^{-2}) \quad . \tag{4.13}$$

A similar analysis yields

$$\mathscr{E}\hat{\sigma}(W) = \sigma(W) + 0(N^{-1}) \tag{4.14}$$

and

$$\operatorname{Var} \hat{\sigma}(W) = \frac{1}{4NB_0^2\sigma^2(W)}\Big\{T_{2,2} - 4\mu_1(W)T_{1,2}$$
$$- 2[\sigma^2(W) + \sigma^2(|P|^2) - \mu_1^2(W)]T_{0,2} + 4\mu_1^2(W)T_{1,1}$$
$$+ 4\mu_1(W)[\sigma^2(W) + \sigma^2(|P|^2) - \mu_1^2(W)]T_{0,1}$$
$$+ [\sigma^2(W) + \sigma^2(|P|^2) - \mu_1^2(W)]^2 T_{0,0}\Big\} + 0(N^{-2}) \quad . \tag{4.15}$$

Thus we see that $\hat{\mu}_1(W)$ and $\hat{\sigma}(W)$ are consistent estimators of $\mu_1(W)$ and $\sigma(W)$, respectively.

If we let $\mathscr{T}$ be a symbolic operator with the property that

$$\mathscr{T}^{\alpha}\mathscr{T}^{\beta} = T_{\alpha,\beta}$$

for nonnegative integers α and β, then we may write (4.13) and (4.15) as

$$\operatorname{Var} \hat{\mu}_1(W) = \frac{1}{NB_0^2}[\mathscr{T}^1 - \mu_1(W)\mathscr{T}^0]^2 + 0(N^{-2})$$

and

$$\text{Var}\,\hat{\sigma}(W) = \frac{1}{4NB_0^2\sigma^2(W)}\{\mathcal{T}^2 - 2\mu_1(W)\mathcal{T}^1$$
$$- [\sigma^2(W) + \sigma^2(|P|^2) - \mu_1^2(W)]\mathcal{T}^0\}^2 + 0\,(N^{-2}) \quad ,$$

respectively.

For concreteness, let us examine a numerical model. We shall assume that W is a Gaussian shaped function,

$$W(f) = \frac{1}{b\sqrt{2\pi}}\exp[-(f-\beta)^2/2b^2] \quad . \tag{4.16}$$

The multiplicative constant in this equation has been chosen as a normalizing factor so that

$$\int_{-\infty}^{\infty} W(f)df = 1 \quad .$$

We also note that the covariance function R (the inverse Fourier transform of W) is complex,

$$R(h) = \exp\,[-\tfrac{1}{2}(2\pi bh)^2]\,\exp\,(i2\pi\beta h) \quad .$$

For convenience let us also assume that p is of Gaussian shape, say

$$p(t) = \left(\frac{2\sqrt{2\pi}}{T}\right)^{1/2} \exp\,(-4\pi^2t^2/T^2)$$

where the parameter $T > 0$ simulates a finite "time window." The multiplicative factor in the equation above also has been chosen as a normalizing factor so that

$$\int_{-\infty}^{\infty} p^2(t)dt = 1 \quad .$$

If, as usual, we denote the Fourier transform of p by P, then

$$|P(f)|^2 = \frac{T}{\sqrt{2\pi}}\exp\,(-\tfrac{1}{2}f^2T^2) \quad .$$

The physical meaning of the parameters b and T should be quite clear. Thus

$$\sigma(W) = b$$

is the population spectral width, and the r.m.s. frequency spread of $|P|^2$ is

$$\sigma(|P|^2) = \frac{1}{T} \quad .$$

It may be referred to as the frequency resolution resulting from the processing of a simulated finite data interval. The parameter

$$\lambda = bT$$

is the ratio of the spectral width of the process to the frequency resolution. Since the "width" of the data interval is proportional to T, and the "correlation time" of the covariance function R is inversely proportional to b, the ratio of these physical quantities is proportional to λ.

Some arithmetic calculations now yield [see (4.12)–(4.15)]

$$\mathscr{E}\hat{\mu}_1(W) = \beta \tag{4.17}$$

$$\text{Var}\,\hat{\mu}_1(W) = \frac{b^2}{2N[1 + (bT)^2]^{1/2}} + 0(N^{-2})$$

$$\mathscr{E}\hat{\sigma}(W) = b + 0(N^{-1})$$

$$\text{Var}\,\hat{\sigma}(W) = \frac{3b^2}{16N[1 + (bT)^2]^{1/2}} + 0(N^{-2}) \quad .$$

The omission of the term $0(N^{-1})$ in (4.17) was intentional. In fact, we shall prove that (4.17) is true for any covariance function (with continuous second derivative) of the form.

$$R(h) = A(h)e^{i2\pi\beta h} \tag{4.18}$$

where β is a real number [see (1.14)]. For the covariance function of (4.18),

$$\mu_1(W) = \beta \quad .$$

To prove (4.17) we see from (4.9) that

$$\hat{\mu}_1(W) = \beta + \sum_{r=0}^{\infty} (-1)^r B_0^{-(r+1)} E_r$$

where

$$E_r = Z_1 Z_0^r - \beta Z_0^{r+1} \quad .$$

Thus we have to show that

$$\mathscr{E} E_r = 0 \quad , \qquad r = 0, 1, \cdots \quad .$$

From (4.8)

$$E_r = \int_{-\infty}^{\infty} \int_{-\infty}^{\infty} (f - \beta)\xi(f)\xi(f_1) \cdots \xi(f_r)\, df\, d\mathbf{f}$$

where

$$\boldsymbol{f}' = \{f_1, \cdots, f_r\}$$

and from (4.3), and the definition of ξ,

$$\xi(f) = c(f) - \gamma(f, f)$$
$$= \frac{1}{N}\sum_{n=1}^{N}\int_{-\infty}^{\infty}\int_{-\infty}^{\infty} p(t)p(s)e^{-i2\pi f(t-s)}[x_n(t)\bar{x}_n(s) - R(t-s)]dtds \quad .$$

Hence

$$E_r = \frac{1}{i4\pi N^{r+1}}\sum_{n=1}^{N}\sum_{n_1=1}^{N}\cdots\sum_{n_r=1}^{N}\int_{-\infty}^{\infty}\int_{-\infty}^{\infty} p^2(t)p^2(t_1)\cdots p^2(t_r)$$
$$\times X(t, \boldsymbol{t})dtd\boldsymbol{t}$$

where

$$X(t, \boldsymbol{t}) = [\bar{x}_n(t)x'_n(t) - x_n(t)\bar{x}'_n(t) - i4\pi\beta|x_n(t)|^2]$$
$$\times \prod_{j=1}^{r}[|x_{n_j}(t_j)|^2 - R(0)] \tag{4.19}$$

and $\boldsymbol{t}' = \{t_1, \cdots, t_r\}$.

Now X is the sum of products of independent terms, each product containing one factor of the form

$$Y = [\bar{x}(t)x'(t) - x(t)\bar{x}'(t) - i4\pi\beta|x(t)|^2]\prod_{j=1}^{q}|x(t_j)|^2 \tag{4.20}$$

for some q, $0 \leqq q \leqq r$. Our problem now reduces to proving that $\mathscr{E}Y = 0$. Toward this end consider

$$x(t)\bar{x}(s)\prod_{j=1}^{q}|x(t_j)|^2 \quad .$$

Then

$$\mathscr{E}x(t)\bar{x}(s)\prod_{j=1}^{q}|x(t_j)|^2 = e^{i2\pi\beta(t-s)}A(t, s, \boldsymbol{t})$$

where $A(t, s, \boldsymbol{t}) = A(s, t, \boldsymbol{t})$ is real and independent of β. If we differentiate with respect to t,

$$\mathscr{E}x'(t)\bar{x}(s)\prod_{j=1}^{q}|x(t_j)|^2 = i2\pi\beta e^{i2\pi\beta(t-s)}A + e^{i2\pi\beta(t-s)}\frac{\partial A}{\partial t} \quad .$$

Thus

$$\mathscr{E}[\bar{x}(t)x'(t) - x(t)\bar{x}'(t)] \prod_{j=1}^{q} |x(t_j)|^2 = i4\pi\beta A(t, t, \boldsymbol{t})$$

and we see [cf. (4.20)] that

$$\mathscr{E}Y = 0 \quad .$$

One may also show that for covariance functions with a linear phase [cf. (4.18)] the statistics (4.13) and (4.15) are independent of β. That is,

$$\operatorname{Var} \hat{\mu}_1(W) = \frac{1}{NB_0^2} T_{1,1} + 0(N^{-2}) \tag{4.21}$$

and

$$\operatorname{Var} \hat{\sigma}(W) = \frac{1}{4NB_0^4\sigma^2(W)}[B_0^2T_{2,2} - 2B_0B_2T_{0,2} + B_2^2T_{0,0}] + 0(N^{-2}) \tag{4.22}$$

where β has been set equal to zero in the B_k and $T_{j,k}$ terms.

To prove this result it is convenient to start with the $T_{j,k}$ statistic [see (4.10)] expressed in the time domain, namely,

$$T_{j,k} = \frac{1}{(i2\pi)^{j+k}} \sum_{\mu=0}^{j} \sum_{\nu=0}^{k} \binom{j}{\mu} \binom{k}{\nu} \int_{-\infty}^{\infty} \int_{-\infty}^{\infty} p(t)p^{(j-\mu)}(t)$$
$$\times\, p(s)p^{(k-\nu)}(s)R^{(\mu)}(t-s)R^{(\nu)}(s-t)dtds \quad ,$$

and assume that

$$\lim_{t\to\pm\infty} p^{(\alpha)}(t) = 0 \quad , \qquad \alpha = 0, 1 \quad .$$

Now if we define $U_{jk\mu\nu}$ as

$$U_{jk\mu\nu} = \int_{-\infty}^{\infty} \int_{-\infty}^{\infty} [p^{(j)}(t)]^2[p^{(k)}(s)]^2 A^{(\mu)}(t-s)A^{(\nu)}(t-s)\, dtds \quad ,$$

then

$$T_{0,0} = U_{0000}$$
$$T_{0,1} = \beta T_{0,0}$$
$$T_{1,1} = \beta^2 T_{0,0} + \frac{1}{8\pi^2}(U_{0011} - U_{0002})$$
$$T_{0,2} = \beta^2 T_{0,0} - \frac{1}{4\pi^2}(U_{0002} - U_{0100})$$
$$T_{1,2} = \beta(2T_{1,1} + T_{0,2}) - 2\beta^3 T_{0,0}$$
$$T_{2,2} = 2\beta T_{1,2} - \beta^4 T_{0,0} + \frac{1}{32\pi^4}(U_{0022} + U_{0004} - 4U_{0102} + 2U_{1100}) \quad .$$

Observe that $U_{jk\mu\nu}$ is independent of β. Thus from (4.13)

$$T_{1,1} - 2\beta T_{0,1} + \beta^2 T_{0,0} = \frac{1}{8\pi^2}(U_{0011} - U_{0002}) = T_{1,1}\Big|_{\beta=0}$$

independent of β, and from (4.15)

$$\operatorname{Var} \hat{\sigma}(W) = \frac{1}{4NB_0^2\sigma^2(W)}\left\{\frac{1}{32\pi^4}(U_{0022} + U_{0004} - 4U_{0102} + 2U_{1100}) + \frac{[\sigma^2(W) + \sigma^2(|P|^2)]}{2\pi^2}(U_{0002} - U_{0100}) + [\sigma^2(W) + \sigma^2(|P|^2)]^2 U_{0000}\right\} + 0(N^{-2}) \tag{4.23}$$

independent of β.

If $W(f)$ is the spectral density function corresponding to the covariance function of (4.18), then $\Omega(f) = W(f + \beta)$ is the spectral density function corresponding to $A(t)$ (and does not depend on β.) But

$$\frac{B_0}{\int_{-\infty}^{\infty} |P(f)|^2 df} = \int_{-\infty}^{\infty} W(g)dg = \int_{-\infty}^{\infty} \Omega(g)dg$$

and

$$\mu_1(W) = \beta \quad , \qquad \mu_1(\Omega) = 0 \quad ,$$

as well as

$$\mu_2(W) - \mu_1^2(W) = \sigma^2(W) = \sigma^2(\Omega) = \mu_2(\Omega) \quad .$$

Thus it is academic whether or not we let $\beta = 0$ in B_0 and $\sigma^2(W)$ of (4.23).

5. Mathematical Justification

Let $\{x(t) | -\infty < t < \infty\}$ be a mean zero stationary complex Gaussian process with continuous positive definite covariance function R and spectral distribution function S. If S is absolutely continuous, then the spectral density $W = S'$ a.e. (Lebesgue measure). We shall further assume that

$$\int_{-\infty}^{\infty} f^4 dS(f) < \infty \quad . \tag{5.1}$$

Then $R^{iv}(0)$ exists (see Corollary 5.2 of Chapter II) and $\{x(t) \mid -\infty < t < \infty\}$ is continuous with probability one (see Theorem 6.5 of Chapter II). Let p be a real, bounded, nonnegative function which is integrable on $(-\infty, \infty)$ with a positive area. Also let p be piecewise differentiable, and let p' be bounded a.e. and integrable on $(-\infty, \infty)$. Thus p is not identically zero; and since R is positive definite, x cannot be identically zero either.

By Theorem 2.6 of Chapter IV, the stochastic integral

$$G(f) = \int_{-\infty}^{\infty} p(t)x(t)e^{-i2\pi ft}dt \tag{5.2}$$

converges both in mean square and with probability one, and the two integrals are equal with probability one. We wish to prove that

$$\int_{-\infty}^{\infty} f^k |G(f)|^2 df$$

exists for $k = 0, 1, 2$. Toward this end we write

$$\int_{-\infty}^{\infty} p^2(t)|x(t)|^2 dt = \int_{-\infty}^{\infty} p^2(t)[|x(t)|^2 - R^2(0)]dt + R^2(0)\int_{-\infty}^{\infty} p^2(t)dt \quad . \tag{5.3}$$

Now $\{|x(t)|^2 - R^2(0) \mid -\infty < t < \infty\}$ is a mean zero process with continuous sample functions and covariance function $r = |R|^2$. By (5.1) and Theorem 5.2 of Chapter II, $r(h)$ is continuous on $(-\infty, \infty)$. Thus by Corollary 2.3 of Chapter IV, and the conditions on p, both integrals on the right-hand side of (5.3) exist. Hence by Parseval's theorem

$$\int_{-\infty}^{\infty} |G(f)|^2 df = \int_{-\infty}^{\infty} p^2(t)|x(t)|^2 dt < \infty \quad .$$

Using Theorem 8.5 and Corollaries 8.2 and 8.7 of Chapter II we also see that

$$\int_{-\infty}^{\infty} f|G(f)|^2 df = \frac{i}{2\pi}\int_{-\infty}^{\infty} p(t)x(t)\frac{d}{dt}[p(t)\bar{x}(t)]dt$$

and

$$\int_{-\infty}^{\infty} f^2|G(f)|^2 df = \frac{1}{4\pi^2}\int_{-\infty}^{\infty} \left|\frac{d}{dt}p(t)x(t)\right|^2 dt$$

exist.

Now define

$$\mu_k(W) = \frac{\int_{-\infty}^{\infty} f^k dS(f)}{\int_{-\infty}^{\infty} dS(f)} \quad .$$

This is consistent with our earlier definition of $\mu_k(W)$ [cf. (0.1) and (0.2); see also (1.1)]. By (5.1), $\mu_k(W)$ certainly exists, at least for $k = 0, 1, 2, 3, 4$.

We now turn to a consideration of the estimators

$$\hat{\mu}_1(W) = \mu_1(c)$$

and

$$\hat{\sigma}(W) = [\sigma^2(c) - \sigma^2(|P|^2)]^{1/2}$$

where

$$c(f) = \frac{1}{N} \sum_{n=1}^{N} |G_n(f)|^2$$

and

$$G_n(f) = \int_{-\infty}^{\infty} p(t) x_n(t) e^{-i2\pi ft} dt \quad . \tag{5.4}$$

As usual, the x_n, $1 \leqq n \leqq N$, are N independent observations on $\{x(t) | -\infty < t < \infty\}$ and P is the Fourier transform of p.

If we define B_k and Z_k as in Section 4, then

$$B_0 = \int_{-\infty}^{\infty} |P(f)|^2 df \int_{-\infty}^{\infty} dS(f) > 0$$

$$B_1 = B_0 \mu_1(W)$$

$$B_2 = B_0[\mu_2(W) + \mu_2(|P|^2)]$$

and

$$\mathscr{E} Z_j Z_k = \frac{1}{N} T_{j,k} \tag{5.5}$$

where

$$T_{j,k} = \int_{-\infty}^{\infty} \int_{-\infty}^{\infty} f^j g^k |\gamma(f, g)|^2 df dg \tag{5.6}$$

and

$$\gamma(f, g) = \int_{-\infty}^{\infty} P(f - h) P(h - g) dS(h) \tag{5.7}$$

[see (4.6)]. We first observe that the $T_{j,k}$ exist for $j, k = 0, 1, 2$. For since $|\gamma(f, g)|^2$ is a covariance function,

$$|\gamma(f, g)|^2 \leqq |\gamma(f, f)| \, |\gamma(g, g)|$$

and hence

$$|T_{j,k}| \leqq \int_{-\infty}^{\infty} |f|^j \gamma(f,f)df \int_{-\infty}^{\infty} |g|^k \gamma(g,g)\,dg \quad .$$

But by (5.7),

$$\int_{-\infty}^{\infty} |f|^j \gamma(f,f)df = \int_{-\infty}^{\infty}\int_{-\infty}^{\infty} |f|^j |P(f-h)|^2 dS(h)df$$
$$= \int_{-\infty}^{\infty}\int_{-\infty}^{\infty} |g+h|^j |P(g)|^2 dS(h)dg \quad .$$

The conditions we have imposed on p imply that $\int_{-\infty}^{\infty} |g|^\alpha |P(g)|^2 dg$ exist for all α, $0 \leqq \alpha \leqq 2$; and by (5.1), $\int_{-\infty}^{\infty} |h|^\alpha dS(h)$ also exists for all α, $0 \leqq \alpha \leqq 2$.

From the definition of $T_{j,k}$ [see (5.6) and (5.5)] we see that

$$\mathscr{E} Z_j Z_k = 0(N^{-1}) \quad .$$

It is not too difficult to show that

$$\mathscr{E} Z_{j_1} Z_{j_2} \cdots Z_{j_{2q}} = 0(N^{-q})$$

and

$$\mathscr{E} Z_{k_1} Z_{k_2} \cdots Z_{k_{2q-1}} = 0(N^{-q}) \quad .$$

We shall need only

$$\mathscr{E} Z_j Z_k Z_l = 0(N^{-2}) \quad , \tag{5.8}$$

$$\mathscr{E} Z_j Z_k Z_l Z_m = 0(N^{-2}) \quad , \tag{5.9}$$

and the fact that the expectation of the product of five or more of the Z_j's is at least $0(N^{-3})$.

After these lengthy but necessary preliminary remarks, let us return to our main theme, namely, the problem of finding the mean and variance of the estimators of the mean frequency and r.m.s. spread. By definition

$$\hat{\mu}_1(W) = \frac{\int_{-\infty}^{\infty} fc(f)df}{\int_{-\infty}^{\infty} c(f)df} = \frac{Z_1 + B_1}{Z_0 + B_0} \quad . \tag{5.10}$$

And if α and $\xi + \alpha$ are nonzero complex numbers, we may recursively apply the arithmetic identity

$$\frac{1}{\xi + \alpha} = \frac{1}{\alpha} - \frac{\xi}{\alpha^2} + \frac{\xi^2}{\alpha^2(\xi + \alpha)} \tag{5.11}$$

to (5.10) (with $\xi = Z_0$ and $\alpha = B_0$) to obtain

$$\hat{\mu}_1(W) = (Z_1 + B_1)\left[\frac{1}{B_0} - \frac{Z_0}{B_0^2} + \frac{Z_0^2}{B_0^3} - \frac{Z_0^3}{B_0^4}\right] + \frac{Z_0^4}{B_0^4}\frac{Z_1 + B_1}{Z_0 + B_0} \quad . \tag{5.12}$$

Since

$$\mathscr{E}(Z_0 - 0)^2 = 0(N^{-1})$$

we see that Z_0 is a consistent estimator of zero. Hence, given any $\varepsilon > 0$ there exists an N_0 such that for all $N > N_0$ we have $Z_0 + B_0$ bounded away from zero with probability $1 - \varepsilon$. For definiteness, say

$$Z_0 + B_0 > \tfrac{1}{2}B_0 > 0 \quad .$$

Let

$$V_0 = Z_0 \qquad \text{if} \quad Z_0 + B_0 > \tfrac{1}{2}B_0$$

$$V_0 = 0 \qquad \text{if} \quad Z_0 + B_0 < \tfrac{1}{2}B_0$$

and let $\tilde{\mu}_1(W)$ be identical with $\hat{\mu}_1(W)$ except that we replace Z_0 by V_0 in the denominator of the last term in (5.12). Then for N sufficiently large, $\tilde{\mu}_1(W)$ will equal $\hat{\mu}_1(W)$ with probability arbitrarily close to one. We also have

$$\mathscr{E}(Z_1 - 0)^2 = 0(N^{-1}) \quad ,$$

which implies that for N sufficiently large

$$Z_1 + B_1 < 1 + B_1$$

with probability arbitrarily close to one. Thus there exists a constant k such that

$$\left|\frac{Z_1 + B_1}{V_0 + B_0}\right| < k$$

with large probability, provided N is large.

Since $\mathscr{E}Z_0^4 = 0(N^{-2})$,

$$\mathscr{E}\frac{Z_0^4}{B_0^4}\frac{Z_1 + B_1}{V_0 + B_0} = 0(N^{-2})$$

and by (5.8) and (5.9)

$$\mathscr{E}\tilde{\mu}_1(W) = \frac{B_1}{B_0} + \frac{B_1}{B_0^3}\mathscr{E}Z_0^2 - \frac{1}{B_0^2}\mathscr{E}Z_0Z_1 + 0(N^{-2})$$

$$= \mu_1(W) - \frac{1}{NB_0^3}(B_0T_{0,1} - B_1T_{0,0}) + 0(N^{-2}) \tag{5.13}$$

and

$$\text{Var}\ \tilde{\mu}_1(W) = \frac{1}{NB_0^4}(B_0^2T_{1,1} - 2B_0B_1T_{0,1} + B_1^2T_{0,0}) + 0(N^{-2}) \quad . \tag{5.14}$$

These last two equations are identical with (4.12) and (4.13), respectively, if we replace $\tilde{\mu}_1$ by $\hat{\mu}_1$.

We now turn to the problem of calculating the statistics of the spread. We have

$$\hat{\sigma}(W) = [\sigma^2(c) - \sigma^2(|P|^2)]^{1/2} = \frac{\sqrt{Z}}{Z_0 + B_0} \tag{5.15}$$

where

$$Z = (Z_2 + B_2)(Z_0 + B_0) - (Z_1 + B_1)^2 - \sigma^2(|P|^2)(Z_0 + B_0)^2 \quad . \tag{5.16}$$

Now if $\eta \geqq 0$ and $\beta > 0$, we may consider the arithmetic identity

$$\sqrt{\eta} = \sqrt{\beta} + \frac{\eta - \beta}{2\sqrt{\beta}} - \frac{(\eta - \beta)^2}{2\sqrt{\beta}\,(\sqrt{\eta} + \sqrt{\beta})^2} \quad . \tag{5.17}$$

Applying this identity and (5.11) to (5.15) (with $\eta = Z, \beta = B_0^2\sigma^2(W), \xi = Z_0, \alpha = B_0$), we obtain

$$\hat{\sigma}(W) = \left\{B_0\sigma(W) + \frac{Z - B_0^2\sigma^2(W)}{2B_0\sigma(W)} - \frac{[Z - B_0^2\sigma^2(W)]^2}{2B_0\sigma(W)[\sqrt{Z} + B_0\sigma(W)]^2}\right\}$$
$$\times \left[\frac{1}{B_0} - \frac{Z_0}{B_0^2} + \frac{Z_0^2}{B_0^2(Z_0 + B_0)}\right] \quad . \tag{5.18}$$

Since

$$\mathscr{E}[Z - B_0^2\sigma^2(W)]^2 = 0(N^{-1}) \tag{5.19}$$

we see that Z is a consistent estimator of $B_0^2\sigma^2(W)$. Hence, given any $\varepsilon > 0$ there exists an N_0 such that for all $N > N_0$ we have $Z \geqq 0$ with probability $1 - \varepsilon$. Let

$$V = Z \quad \text{if} \quad Z \geqq 0$$
$$V = 0 \quad \text{if} \quad Z < 0 \quad ,$$

and let $\tilde{\sigma}(W)$ be identical with $\hat{\sigma}(W)$ except that we replace $\sqrt{Z}$ by $\sqrt{V}$ in the denominator of the third term in braces in (5.18) and replace Z_0 by V_0 in the denominator of the third term in large brackets in (5.18). Then for N sufficiently large, $\tilde{\sigma}(W)$ will equal $\hat{\sigma}(W)$ with probability arbitrarily close to one.

Now from

$$\mathscr{E}[Z - B_0^2\sigma^2(W)] = 0(N^{-1}) = \mathscr{E}[Z - B_0^2\sigma^2(W)]^2 \quad ,$$

$$\mathscr{E}[Z - B_0^2\sigma^2(W)]^3 = 0(N^{-2}) = \mathscr{E}[Z - B_0^2\sigma^2(W)]^4 \quad ,$$

and

$$\frac{[Z - B_0^2\sigma^2(W)]^2}{[\sqrt{V} + B_0\sigma(W)]^2} \leqq \frac{[Z - B_0^2\sigma^2(W)]^2}{B_0^2\sigma^2(W)}$$

we have

$$\mathscr{E}\,\frac{[Z - B_0^2\sigma^2(W)]^2}{[\sqrt{V} + B_0\sigma(W)]^2} = 0(N^{-1}) \quad ,$$

as well as

$$\mathscr{E}\,\tilde{\sigma}(W) = \sigma(W) + 0(N^{-1}) \tag{5.20}$$

and

$$\begin{aligned} \operatorname{Var}\tilde{\sigma}(W) = \frac{1}{4NB_0^6\sigma^2(W)}\,[&B_0^4T_{2,2} - 4B_0^3B_1T_{1,2} + 4B_0^2B_1^2T_{1,1} \\ &- 2B_0^2(B_0B_2 - 2B_1^2)T_{0,2} \\ &+ 4B_0B_1(B_0B_2 - 2B_1^2)T_{0,1} \\ &+ (B_0B_2 - 2B_1^2)^2T_{0,0}] + 0(N^{-2}) \quad . \end{aligned} \tag{5.21}$$

These last two equations are identical with (4.14) and (4.15), respectively, if we replace $\tilde{\sigma}$ by $\hat{\sigma}$.

If we define

$$\hat{\Sigma}(W) = \sigma^2(c) - \sigma^2(|P|^2) \quad , \tag{5.22}$$

an estimator of the *square* of the spread, then

$$\begin{aligned} \mathscr{E}\,\hat{\Sigma}(W) = \sigma^2(W) - \frac{1}{NB_0^4}\,[&B_0^2T_{1,1} + B_0^2T_{0,2} - 4B_0B_1T_{0,1} \\ &- (B_0B_2 - 3B_1^2)T_{0,0}] + 0(N^{-2}) \end{aligned} \tag{5.23}$$

and

$$\operatorname{Var}\hat{\Sigma}(W) = \frac{1}{NB_0^6}\,[B_0^4T_{2,2} - 4B_0^3B_1T_{1,2} + 4B_0^2B_1^2T_{1,1}$$

$$- 2B_0^2(B_0B_2 - 2B_1^2)T_{0,2} + 4B_0B_1(B_0B_2 - 2B_1^2)T_{0,1}$$
$$+ (B_0B_2 - 2B_1^2)^2T_{0,0}] + 0(N^{-2}) \qquad (5.24)$$

where $\tilde{\Sigma}$ is defined in the expected fashion.

Now if Σ and σ are random variable related by the equation

$$\Sigma = \sigma^2 \quad ,$$

then formally, to first-order approximation,

$$\text{Var } \Sigma = \left(\frac{d\Sigma}{d\sigma}\right)^2_{\sigma=\mathscr{E}\sigma} \text{Var } \sigma \quad .$$

If we interpret Σ as $\tilde{\Sigma}$ and σ as $\tilde{\sigma}$, then $\mathscr{E}\tilde{\sigma}(W) = \sigma(W)$ and

$$\text{Var } \tilde{\Sigma}(W) = 4\sigma^2(W) \text{ Var } \tilde{\sigma}(W) \quad .$$

But this is precisely the relationship that exists (to order N^{-2}) between Var $\tilde{\Sigma}(W)$ and Var $\tilde{\sigma}(W)$ [see (5.24) and (5.21)].

VII

HYPOTHESIS TESTING

0. Introduction

In Chapter V we considered certain topics related to the problem of estimating parameters of signals embedded in complex noise, and in Chapter VI we considered the problem of estimating spectral moments. Now the covariance function R of a stochastic process reveals much information about the process. In fact we know that in the Gaussian case it tells the whole story. Suppose $\{y(t) \,|\, t \in T\}$ is a mean zero complex process and we make N sets of measurements at k distinct time points $t_1, \cdots, t_k$ in T. Then if $\boldsymbol{y}_n' = \{y_n(t_1), \cdots, y_n(t_k)\}$, $1 \leqq n \leqq N$, the covariance matrix $\boldsymbol{R} = \mathscr{E}\boldsymbol{y}_n\bar{\boldsymbol{y}}_n'$ of $\boldsymbol{y}_n$ is a $k \times k$ matrix. This matrix contains considerable information. Suppose, for example (see Section 5 of Chapter V and Chapter VI) that $k = 2$ and that the $\{y(t)\}$ process is wide-sense stationary with covariance function of the form

$$R(h) = a \exp\left[-\tfrac{1}{2}(2\pi bh)^2\right] \exp(i2\pi ch) \quad .$$

Then a may be interpreted as the total power, b is the spread, and c is the mean frequency.

Now suppose we make observations on a process whose covariance matrix is either $\boldsymbol{R}_0$ or $\boldsymbol{R}_1$ and we wish to make a decision, based on the vector measurements, as to which process we are observing. We may formulate this problem as an hypothesis testing problem. That is, we wish to test the simple hypothesis

$$\mathscr{H}_0 \quad : \quad \boldsymbol{R} = \boldsymbol{R}_0$$

versus the simple alternative hypothesis

$$\mathscr{H}_1 \quad : \quad \boldsymbol{R} = \boldsymbol{R}_1 \quad .$$

We shall analyze this situation in the present chapter for the important case

$k = 2$. (In engineering literature one frequently refers to the case where the vectors are two-dimensional as the "pulse-pair" case.)

In Section 1 we derive the likelihood ratio statistic q for testing $\mathscr{H}_0$ versus $\mathscr{H}_1$. It is seen that a certain matrix $\boldsymbol{\Sigma}$ that appears in the formulation can be definite, indefinite, or semidefinite. This leads to three cases for which we must compute the probability density function of q. We accomplish this in Sections 2, 3, and 4. Finally, in Section 5, we compute the moments of the statistic, and make some general comments on the computation of the error probabilities.

1. The Likelihood Ratio Statistic

We recall a few results from hypothesis testing theory (see, for example, [HC]). Suppose we are testing an hypothesis $\mathscr{H}_0$ against an alternative hypothesis $\mathscr{H}_1$. Then there is always the risk of making an incorrect decision. That is, we accept $\mathscr{H}_1$ when $\mathscr{H}_0$ is true, or we accept $\mathscr{H}_0$ when $\mathscr{H}_1$ is true. It is customary to refer to these errors as Type I and Type II errors, respectively. More explicitly, we define:

$$\begin{aligned}\alpha &= \text{probability of committing a Type I error}\\ &= P[\text{Accept} \quad \mathscr{H}_1 | \mathscr{H}_0 \quad \text{true}]\end{aligned}$$

and

$$\begin{aligned}\beta &= \text{probability of committing a Type II error}\\ &= P[\text{Accept} \quad \mathscr{H}_0 | \mathscr{H}_1 \quad \text{true}].\end{aligned}$$

One often calls α the *missed signal probability* and $1 - \alpha$ the *detection probability*. Also, β is referred to as the *false alarm probability* and $1 - \beta$ is called the *power of the test*.

To choose a criterion for testing an hypothesis $\mathscr{H}_0$ against an alternative hypothesis $\mathscr{H}_1$ is to partition the sample space into two sets—a region of acceptance for $\mathscr{H}_0$ and a region of rejection for $\mathscr{H}_0$. The region of rejection for $\mathscr{H}_0$ is called the critical region. The probability that a sample point will fall into the critical region when the simple hypothesis $\mathscr{H}_0$ is true is often referred to as the size of the critical region, namely, α. In the case $\mathscr{H}_0$ and $\mathscr{H}_1$ are both simple, a critical region of size α is said to be most powerful if the probability of committing a Type II error is minimized.

Consider now the problem of testing the simple hypothesis

$$\mathscr{H}_0 \quad : \quad \gamma = \gamma_0$$

versus the simple alternative hypothesis

$$\mathscr{H}_1 \quad : \quad \gamma = \gamma_1 \quad ,$$

where γ is a parameter of the density function $f(x; \gamma)$. Let $\boldsymbol{x}' = \{x_1, \cdots, x_N\}$ be an N-dimensional sample from $f(x; \gamma)$. Then

$$L_0 = L(\boldsymbol{x}; \gamma | \mathscr{H}_0) = L(\boldsymbol{x}; \gamma_0) = \prod_{n=1}^{N} f(x_n; \gamma_0)$$

is the likelihood function of the sample from this distribution when $\gamma = \gamma_0$, that is, when the hypothesis $\mathscr{H}_0$ is true. Similarly,

$$L_1 = L(\boldsymbol{x}; \gamma | \mathscr{H}_1) = L(\boldsymbol{x}; \gamma_1) = \prod_{n=1}^{N} f(x_n; \gamma_1)$$

is the likelihood function when $\gamma = \gamma_1$. Then the Neyman–Pearson lemma states that if $\mathscr{C}$ is a critical region of size α, and if there exists a constant $\lambda > 0$ such that

$$\frac{L_0}{L_1} \leqq \lambda \qquad \text{inside} \quad \mathscr{C} \tag{1.1}$$

and

$$\frac{L_0}{L_1} \geqq \lambda \qquad \text{outside} \quad \mathscr{C} \quad , \tag{1.2}$$

then $\mathscr{C}$ is a most powerful critical region for testing $\gamma = \gamma_0$ against $\gamma = \gamma_1$.

Now suppose we can determine a statistic $q = q(\boldsymbol{x})$ from (1.1) and (1.2) such that

$$q \geqq K \qquad \text{inside} \quad \mathscr{C}$$

$$q \leqq K \qquad \text{outisde} \quad \mathscr{C}$$

where K depends on λ and perhaps the population parameters, but not on the observations $\boldsymbol{x}$. Then given an α we can determine a value of K such that

$$\alpha = P[q \geqq K | \mathscr{H}_0 \quad \text{true}]$$

and hence with this value of K, we can determine β by the equation

$$1 - \beta = P[q \geqq K | \mathscr{H}_1 \quad \text{true}] \quad .$$

The difficult part, in general, is to find the probability density function F of q under the hypotheses $\mathscr{H}_0$ and $\mathscr{H}_1$. If for example $F(q | \mathscr{H}_0)$ is the desired distribution assuming $\mathscr{H}_0$ true, then

$$\alpha = \int_K^\infty F(q | \mathscr{H}_0) dq = P[\text{Accept} \quad \mathscr{H}_1 | \mathscr{H}_0 \quad \text{true}] \quad . \tag{1.3}$$

Similarly, if $F(q|\mathscr{H}_1)$ is the distribution of q under the hypothesis $\mathscr{H}_1$, then

$$1 - \beta = \int_K^\infty F(q|\mathscr{H}_1)dq = P[\text{Accept} \quad \mathscr{H}_1|\mathscr{H}_1 \quad \text{true}] \ . \qquad (1.4)$$

Now let $\{y(t)| -\infty < t < \infty\}$ be a mean zero complex stationary Gaussian process with covariance function R. Let t_1 and t_2 be two distinct time points, and let

$$\boldsymbol{y}_n' = \{y_n(t_1), y_n(t_2)\} \quad , \qquad 1 \leqq n \leqq N \ ,$$

be N (independent) vector observations. Then

$$\mathscr{E}\boldsymbol{y}_n = \boldsymbol{0}$$

and

$$\mathscr{E}\boldsymbol{y}_n\bar{\boldsymbol{y}}_n' = \boldsymbol{R} \qquad \text{(independent of } n\text{)}$$

where we shall assume that the 2×2 covariance matrix

$$\boldsymbol{R} = \begin{bmatrix} R(0) & R(h) \\ \bar{R}(h) & R(0) \end{bmatrix} \quad , \qquad h = t_1 - t_2 \neq 0 \quad ,$$

is positive definite.

Consider the hypothesis

$$\mathscr{H}_0 \quad : \qquad \boldsymbol{R} = \boldsymbol{R}_0$$

and the alternative hypothesis

$$\mathscr{H}_1 \quad : \qquad \boldsymbol{R} = \boldsymbol{R}_1 \quad .$$

The density function of $\boldsymbol{y}_n$ under $\mathscr{H}_0$ is

$$f(\boldsymbol{y}_n|\mathscr{H}_0) = \frac{1}{\pi^2 \det \boldsymbol{R}_0} \exp(-\bar{\boldsymbol{y}}_n'\boldsymbol{R}_0^{-1}\boldsymbol{y}_n)$$

and the density function of $\boldsymbol{y}_n$ under $\mathscr{H}_1$ is

$$f(\boldsymbol{y}_n|\mathscr{H}_1) = \frac{1}{\pi^2 \det \boldsymbol{R}_1} \exp(-\bar{\boldsymbol{y}}_n'\boldsymbol{R}_1^{-1}\boldsymbol{y}_n) \quad .$$

Thus the likelihood function L under $\mathscr{H}_0$ is

$$L_0 = L(\boldsymbol{y}_1, \cdots, \boldsymbol{y}_N|\mathscr{H}_0) = \frac{1}{\pi^{2N} (\det \boldsymbol{R}_0)^N} \exp\left(-\sum_{n=1}^N \bar{\boldsymbol{y}}_n' \boldsymbol{R}_0^{-1}\boldsymbol{y}_n\right)$$

and the likelihood function L under $\mathscr{H}_1$ is

$$L_1 = L(\boldsymbol{y}_1, \cdots, \boldsymbol{y}_N | \mathscr{H}_1) = \frac{1}{\pi^{2N}(\det \boldsymbol{R}_1)^N} \exp\left(-\sum_{n=1}^{N} \bar{\boldsymbol{y}}_n' \boldsymbol{R}_1^{-1} \boldsymbol{y}_n\right) \quad .$$

Thus we may form the likelihood ratio

$$\frac{L_0}{L_1} = \frac{(\det \boldsymbol{R}_1)^N}{(\det \boldsymbol{R}_0)^N} \exp\left[-\sum_{n=1}^{N} \bar{\boldsymbol{y}}_n'(\boldsymbol{R}_0^{-1} - \boldsymbol{R}_1^{-1})\,\boldsymbol{y}_n\right] \leqq \lambda \quad .$$

Now if we take the logarithm of the likelihood ratio, there results

$$\sum_{n=1}^{N} \bar{\boldsymbol{y}}_n' \boldsymbol{\Sigma} \boldsymbol{y}_n \geqq K$$

where

$$\boldsymbol{\Sigma} = \boldsymbol{R}_0^{-1} - \boldsymbol{R}_1^{-1} \tag{1.5}$$

and

$$K = N \log \det \boldsymbol{R}_1 - N \log \det \boldsymbol{R}_0 - \log \lambda \quad .$$

Let

$$q = \sum_{n=1}^{N} \bar{\boldsymbol{y}}_n' \boldsymbol{\Sigma} \boldsymbol{y}_n \quad . \tag{1.6}$$

Then the probability α of committing a Type I error, that is, of accepting $\boldsymbol{R} = \boldsymbol{R}_1$ when $\boldsymbol{R} = \boldsymbol{R}_0$ is true, is

$$\alpha = P[q \geqq K | \mathscr{H}_0 \quad \text{true}]$$

and the probability of accepting $\boldsymbol{R} = \boldsymbol{R}_1$ when $\boldsymbol{R} = \boldsymbol{R}_1$ is indeed true, that is, the probability of making a correct decision, is

$$1 - \beta = P[q \geqq K | \mathscr{H}_1 \quad \text{true}] \quad .$$

Our problem now reduces to one of finding the probability density function F of q given $\mathscr{H}_0$ or $\mathscr{H}_1$. That is, we must compute $F(q|\mathscr{H}_0)$ and $F(q|\mathscr{H}_1)$. Once this has been done, α, β, and K are determined by (1.3) and (1.4).

Before proceeding further, however, let us look more closely at the matrix $\boldsymbol{\Sigma}$ [cf. (1.5)]. First we note that since $\{y(t)\}$ is stationary, $\boldsymbol{\Sigma}$ has equal diagonal elements:

$$\boldsymbol{\Sigma} = \begin{bmatrix} \Sigma_0 & \Sigma_1 \\ \bar{\Sigma}_1 & \Sigma_0 \end{bmatrix} \quad .$$

Except for the trivial case $\Sigma_0 = |\Sigma_1| = 0$ (which implies that $\boldsymbol{R}_0$ and $\boldsymbol{R}_1$ are identical) three cases arise:

$$|\Sigma_0| > |\Sigma_1|$$
$$|\Sigma_0| < |\Sigma_1|$$
$$|\Sigma_0| = |\Sigma_1| > 0$$

corresponding, respectively, to the condition that $\boldsymbol{\Sigma}$ be definite, indefinite, or semidefinite. We treat all three cases (see Sections 2, 3, and 4).

In any case

$$\begin{aligned}\det \boldsymbol{\Sigma} &= \det(\boldsymbol{R}_0^{-1} - \boldsymbol{R}_1^{-1}) = \det \boldsymbol{R}_0^{-1} \boldsymbol{R}_0(\boldsymbol{R}_0^{-1} - \boldsymbol{R}_1^{-1})\boldsymbol{R}_1\boldsymbol{R}_1^{-1} \\ &= (\det \boldsymbol{R}_0)^{-1} \det(\boldsymbol{R}_1 - \boldsymbol{R}_0) (\det \boldsymbol{R}_1)^{-1} \\ &= \frac{S_d}{S_0 S_1}\end{aligned} \tag{1.7}$$

where

$$\begin{aligned}S_0 &= \det \boldsymbol{R}_0 \\ S_1 &= \det \boldsymbol{R}_1 \\ S_d &= \det(\boldsymbol{R}_1 - \boldsymbol{R}_0) \quad .\end{aligned} \tag{1.8}$$

Note that $\det (\boldsymbol{R}_1 - \boldsymbol{R}_0) = \det(\boldsymbol{R}_0 - \boldsymbol{R}_1)$ since two is an even integer. If $\boldsymbol{\Sigma}$ is definite, $S_d > 0$; if $\boldsymbol{\Sigma}$ is indefinite, $S_d < 0$; and if $\boldsymbol{\Sigma}$ is semidefinite, $S_d = 0$.

We may also write

$$S_0 + S_1 - S_d = S_1 S_{01} \tag{1.9}$$

where

$$S_{01} = \operatorname{tr} \boldsymbol{R}_0 \boldsymbol{R}_1^{-1} \quad . \tag{1.10}$$

Equation (1.9) follows from the identities

$$\det(\boldsymbol{R}_0 + \boldsymbol{R}_1) + \det(\boldsymbol{R}_0 - \boldsymbol{R}_1) = 2(\det \boldsymbol{R}_0 + \det \boldsymbol{R}_1) \tag{1.11}$$

and

$$\det(\boldsymbol{R}_0 + \boldsymbol{R}_1) - \det(\boldsymbol{R}_0 - \boldsymbol{R}_1) = 2(\det \boldsymbol{R}_1) (\operatorname{tr} \boldsymbol{R}_0 \boldsymbol{R}_1^{-1}) \quad . \tag{1.12}$$

[In fact, (1.11) and (1.12) are true for *any* 2×2 matrices $\boldsymbol{R}_0$ and $\boldsymbol{R}_1$, provided only that in (1.12) the matrix $\boldsymbol{R}_1$ be nonsingular.]

2. The Definite Case

Let $\boldsymbol{y}_n$, $1 \leq n \leq N$, be N independent two-dimensional complex Gaussian vectors with means zero and common positive definite covariance matrix $\boldsymbol{R}$. We wish to find the distribution of

$$q = \sum_{n=1}^{N} \bar{\boldsymbol{y}}_n' \boldsymbol{\Sigma} \boldsymbol{y}_n$$

under the hypothesis $\boldsymbol{R} = \boldsymbol{R}_0$ and under the alternative hypothesis $\boldsymbol{R} = \boldsymbol{R}_1$ where $\boldsymbol{\Sigma} = \boldsymbol{R}_0^{-1} - \boldsymbol{R}_1^{-1}$ is definite.

Let us assume that $\boldsymbol{\Sigma}$ is *positive* definite. This is equivalent to letting Σ_0 be positive. If $\Sigma_0 < 0$, that is, if $\boldsymbol{\Sigma}$ is *negative* definite, we may let

$$q^* = \sum_{n=1}^{N} \bar{\boldsymbol{y}}_n'(-\boldsymbol{\Sigma})\boldsymbol{y}_n$$

be our test statistic. (Note that $-\boldsymbol{\Sigma}$ is *positive* definite.) Then

$$\alpha^* = P[q^* \leqq K^* | \mathscr{H}_0 \quad \text{true}] = \int_{-\infty}^{K^*} F^*(q^* | \mathscr{H}_0) dq^*$$

where F^* is the density function of q^*.

Since $\boldsymbol{\Sigma}$ is positive definite, there exists a nonsingular matrix $\boldsymbol{P}$ such that

$$\bar{\boldsymbol{P}}' \boldsymbol{\Sigma} \boldsymbol{P} = \boldsymbol{I} \tag{2.1}$$

where $\boldsymbol{I}$ is the 2×2 identity matrix. Let

$$\boldsymbol{z}_n = \boldsymbol{P}^{-1} \boldsymbol{y}_n \quad .$$

Then

$$\bar{\boldsymbol{y}}_n' \boldsymbol{\Sigma} \boldsymbol{y}_n = \bar{\boldsymbol{z}}_n' \bar{\boldsymbol{P}}' \boldsymbol{\Sigma} \boldsymbol{P} \boldsymbol{z}_n = \bar{\boldsymbol{z}}_n' \boldsymbol{z}_n = |\boldsymbol{z}_n|^2$$

and

$$q = \sum_{n=1}^{N} |\boldsymbol{z}_n|^2$$

where

$$\mathscr{E} \boldsymbol{z}_n = \boldsymbol{P}^{-1} \mathscr{E} \boldsymbol{y}_n = \boldsymbol{0}$$

and

$$\mathscr{E} \boldsymbol{z}_n \bar{\boldsymbol{z}}_n' = \mathscr{E} \boldsymbol{P}^{-1} \boldsymbol{y}_n \bar{\boldsymbol{y}}_n' \bar{\boldsymbol{P}}'^{-1} = \boldsymbol{P}^{-1} \boldsymbol{R} \bar{\boldsymbol{P}}'^{-1} = (\text{say}) \quad \boldsymbol{M} \quad ,$$

is positive definite.

We shall compute the frequency function F of q by first calculating its characteristic function. Since the $|\boldsymbol{z}_n|^2$, $1 \leqq n \leqq N$, are independent and identically distributed, they have a common characteristic function, say ϕ. Thus χ, the characteristic function of q, is given by

$$\chi = \phi^N$$

and the density function F of q is the Fourier transform of χ, viz.:

$$F(q) = \frac{1}{2\pi}\int_{-\infty}^{\infty} \chi(t)e^{-iqt}dt \quad .$$

We begin by computing ϕ. Let $\boldsymbol{z}' = \{z_1, z_2\}$ be normal $(\boldsymbol{0}, \boldsymbol{M})$. Then the joint probability density function of $r_1 = |z_1|$ and $r_2 = |z_2|$ is [see (3.11) of Chapter III]

$$\begin{aligned} g(r_1, r_2) &= \frac{4r_1r_2}{\det \boldsymbol{M}} \exp[-(\mu r_1^2 + \nu r_2^2)]\, I_0(2|\lambda|r_1r_2) \quad , \qquad r_1, r_2 \geqq 0 \quad , \\ &= 0 \quad , \qquad \text{otherwise} \quad , \end{aligned} \tag{2.2}$$

where

$$\boldsymbol{M}^{-1} = \begin{bmatrix} \mu & \lambda \\ \bar{\lambda} & \nu \end{bmatrix} \quad . \tag{2.3}$$

The characteristic function ϕ of

$$|\boldsymbol{z}|^2 = |z_1|^2 + |z_2|^2 = r_1^2 + r_2^2$$

is

$$\phi(t) = \mathscr{E} \exp[i(r_1^2 + r_2^2)t] = \int_{-\infty}^{\infty}\int_{-\infty}^{\infty} \exp[i(r_1^2 + r_2^2)t]\, g(r_1, r_2)dr_1dr_2 \quad .$$

If we expand the modified Bessel function I_0 in g and integrate term by term, we obtain

$$\begin{aligned} \phi(t) &= \frac{1}{\det \boldsymbol{M}} \frac{1}{(\mu - it)(\nu - it) - |\lambda|^2} \\ &= \frac{\det \boldsymbol{M}^{-1}}{[\frac{1}{2}(A + B) - it][\frac{1}{2}(A - B) - it]} \end{aligned} \tag{2.4}$$

where

$$A = \mu + \nu = \operatorname{tr} \boldsymbol{M}^{-1}$$

$$B = [(\mu - \nu)^2 + 4|\lambda|^2]^{1/2} = [(\operatorname{tr} \boldsymbol{M}^{-1})^2 - 4 \det \boldsymbol{M}^{-1}]^{1/2} \quad .$$

Note that

$$A > B \geqq 0 \quad . \tag{2.5}$$

For Re $a > 0$, Re $b > 0$, Re $\beta > 0$ we have the Fourier transform formula

$$\frac{1}{2\pi}\int_{-\infty}^{\infty}(a-it)^{-\beta}(b-it)^{-\beta}\exp(-iqt)\,dt$$

$$= \frac{\pi^{1/2}}{\Gamma(\beta)}\left(\frac{q}{a-b}\right)^{\beta-1/2}\exp[-\tfrac{1}{2}(a+b)q]\, I_{\beta-1/2}(\tfrac{1}{2}(a-b)q) \quad , \qquad q>0 \quad ,$$

$$= 0 \quad , \qquad q<0 \quad ,$$

where $I_{\beta-1/2}$ is the modified Bessel function of the first kind and order $\beta - \frac{1}{2}$. Thus

$$\begin{aligned} F(q) &= \frac{1}{2\pi}\int_{-\infty}^{\infty}\chi(t)e^{-iqt}dt = \frac{1}{2\pi}\int_{-\infty}^{\infty}\phi^N(t)e^{-iqt}dt \\ &= \frac{\pi^{1/2}q^{N-1/2}}{(N-1)!\,(\det \boldsymbol{M})^N B^{N-1/2}}\,e^{-Aq/2}\, I_{N-1/2}(\tfrac{1}{2}Bq) \quad , \qquad q>0 \quad , \\ &= 0 \quad , \qquad q<0 \quad . \end{aligned} \tag{2.6}$$

Under the hypothesis $\mathscr{H}_0$, $\boldsymbol{R} = \boldsymbol{R}_0$ and $\boldsymbol{M} = \boldsymbol{M}_0 = \boldsymbol{P}^{-1}\boldsymbol{R}_0\bar{\boldsymbol{P}}'^{-1}$. We shall write

$$\begin{aligned} A_0 &= \operatorname{tr} \boldsymbol{M}_0^{-1} \\ B_0 &= [(\operatorname{tr} \boldsymbol{M}_0^{-1})^2 - 4\det \boldsymbol{M}_0^{-1}]^{1/2} \quad . \end{aligned}$$

Under the alternative hypothesis $\mathscr{H}_1$, $\boldsymbol{R} = \boldsymbol{R}_1$ and $\boldsymbol{M} = \boldsymbol{M}_1 = \boldsymbol{P}^{-1}\boldsymbol{R}_1\bar{\boldsymbol{P}}'^{-1}$. We define A_1 and B_1 in the expected fashion. Now the diagonalizing matrix $\boldsymbol{P}$ is not unique [cf. (2.1)]. Therefore det $\boldsymbol{M}_0$, A_0, B_0, det $\boldsymbol{M}_1$, A_1, B_1 are independent of $\boldsymbol{P}$. Let us explicitly demonstrate this fact by writing these quantities in terms of $S_0 = \det \boldsymbol{R}_0$, $S_1 = \det \boldsymbol{R}_1$, and $S_d = \det(\boldsymbol{R}_1 - \boldsymbol{R}_0)$ [see (1.8)].

Now

$$\det \boldsymbol{M}_0 = \det \boldsymbol{P}^{-1}\boldsymbol{R}_0\bar{\boldsymbol{P}}'^{-1} = (\det \boldsymbol{\Sigma})\,(\det \boldsymbol{R}_0) = \frac{S_d}{S_1} \tag{2.7}$$

since $\boldsymbol{\Sigma}^{-1} = \boldsymbol{P}\bar{\boldsymbol{P}}'$ [see (2.1) and (1.7)]. Also

$$\begin{aligned} A_0 &= \operatorname{tr} \boldsymbol{M}_0^{-1} = \operatorname{tr} \boldsymbol{\Sigma}^{-1}\boldsymbol{R}_0^{-1} = \operatorname{tr}(\boldsymbol{R}_0^{-1} - \boldsymbol{R}_1^{-1})^{-1}\boldsymbol{R}_0^{-1} \\ &= \operatorname{tr}[\boldsymbol{R}_0(\boldsymbol{R}_0^{-1} - \boldsymbol{R}_1^{-1})]^{-1} = \operatorname{tr}(\boldsymbol{I} - \boldsymbol{R}_0\boldsymbol{R}_1^{-1})^{-1} \quad . \end{aligned}$$

We recall that if $\boldsymbol{\Phi}$ is any nonsingular 2×2 matrix, then

$$\operatorname{tr} \boldsymbol{\Phi} = (\det \boldsymbol{\Phi})\,(\operatorname{tr} \boldsymbol{\Phi}^{-1}) \quad . \tag{2.8}$$

Thus

$$A_0 = \det{}^{-1}(\boldsymbol{I} - \boldsymbol{R}_0\boldsymbol{R}_1^{-1})\operatorname{tr}(\boldsymbol{I} - \boldsymbol{R}_0\boldsymbol{R}_1^{-1}) \quad .$$

But

$$\det (\boldsymbol{I} - \boldsymbol{R}_0\boldsymbol{R}_1^{-1}) = \det (\boldsymbol{I} - \boldsymbol{R}_0\boldsymbol{R}_1^{-1})\, \boldsymbol{R}_1\boldsymbol{R}_1^{-1}$$
$$= \det (\boldsymbol{R}_1 - \boldsymbol{R}_0)\boldsymbol{R}_1^{-1}$$
$$= \frac{S_d}{S_1}$$

and from (1.9)

$$(\det \boldsymbol{R}_1)\,(\mathrm{tr}\ \boldsymbol{R}_0\boldsymbol{R}_1^{-1}) = \det \boldsymbol{R}_0 + \det \boldsymbol{R}_1 - \det (\boldsymbol{R}_1 - \boldsymbol{R}_0) \quad .$$

Thus

$$A_0 = \frac{S_1 - S_0 + S_d}{S_d} \quad . \tag{2.9}$$

Also

$$B_0 = [(\mathrm{tr}\ \boldsymbol{M}_0^{-1})^2 - 4 \det \boldsymbol{M}_0^{-1}]^{1/2}$$
$$= (A_0^2 - 4S_1S_d^{-1})^{1/2}$$
$$= \frac{1}{S_d}[(S_0 + S_1 - S_d)^2 - 4S_0S_1]^{1/2}$$
$$= \frac{1}{S_d}(S_1^2S_{01}^2 - 4S_0S_1)^{1/2} \quad .$$

Similarly,

$$\det \boldsymbol{M}_1 = \frac{S_d}{S_0}$$

$$A_1 = \frac{S_1 - S_0 - S_d}{S_d} \tag{2.10}$$

(and hence $A_0 - A_1 = 2$). Also,

$$B_1 = B_0 \quad .$$

We shall call their common value B.

To summarize, then, under the hypothesis $\mathscr{H}_0$, the frequency function of q is

$$F(q\,|\,\mathscr{H}_0) = \frac{\pi^{1/2}S_1^Nq^{N-1/2}}{(N-1)!\ S_d^NB^{N-1/2}}\exp(-\tfrac{1}{2}A_0q)\ I_{N-1/2}(\tfrac{1}{2}Bq) \quad , \qquad q > 0 \quad ,$$
$$= 0 \quad , \qquad q < 0 \quad , \tag{2.11}$$

and under the alternative hypothesis $\mathscr{H}_1$, the frequency function of q is

$$F(q|\mathscr{H}_1) = \frac{\pi^{1/2} S_0^N q^{N-1/2}}{(N-1)!\ S_d^N B^{N-1/2}} \exp(-\tfrac{1}{2}A_1 q)\ I_{N-1/2}(\tfrac{1}{2}Bq)\ , \qquad q > 0\ ,$$
$$= 0\ , \qquad q < 0\ , \tag{2.12}$$

where A_0 and A_1 are given by (2.9) and (2.10) while

$$B = \frac{1}{S_d}[(S_0 + S_1 - S_d)^2 - 4S_0S_1]^{1/2}\ . \tag{2.13}$$

3. The Indefinite Case

Let $\boldsymbol{y}_n$, $1 \leqq n \leqq N$, be N independent two-dimensional complex Gaussian vectors with means zero and common positive definite covariance matrix $\boldsymbol{R}$. We wish to find the distribution of

$$q = \sum_{n=1}^{N} \bar{\boldsymbol{y}}_n' \boldsymbol{\Sigma} \boldsymbol{y}_n$$

under the hypothesis $\boldsymbol{R} = \boldsymbol{R}_0$ and under the alternative hypothesis $\boldsymbol{R} = \boldsymbol{R}_1$ where $\boldsymbol{\Sigma} = \boldsymbol{R}_0^{-1} - \boldsymbol{R}_1^{-1}$ is indefinite.

Since the matrix $\boldsymbol{\Sigma}$ is assumed indefinite, there exists a nonsingular matrix $\boldsymbol{P}$ such that

$$\bar{\boldsymbol{P}}'\boldsymbol{\Sigma}\boldsymbol{P} = \boldsymbol{J} \tag{3.1}$$

where

$$\boldsymbol{J} = \begin{bmatrix} 1 & 0 \\ 0 & -1 \end{bmatrix}\ . \tag{3.2}$$

Let

$$\boldsymbol{z}_n = \boldsymbol{P}^{-1}\boldsymbol{y}_n\ .$$

Then

$$\bar{\boldsymbol{y}}_n'\boldsymbol{\Sigma}\boldsymbol{y}_n = \bar{\boldsymbol{z}}_n'\bar{\boldsymbol{P}}'\boldsymbol{\Sigma}\boldsymbol{P}\boldsymbol{z}_n = |z_{n,1}|^2 - |z_{n,2}|^2$$

where $\boldsymbol{z}_n' = \{z_{n,1}, z_{n,2}\}$, and

$$q = \sum_{n=1}^{N} (|z_{n,1}|^2 - |z_{n,2}|^2)\ .$$

Also

$$\mathscr{E} z_n = 0$$

and

$$\mathscr{E} z_n \bar{z}_n' = P^{-1} R \bar{P}'^{-1} = \text{(say)} \quad M$$

where M is positive definite.

As in Section 2, we wish to compute the frequency function F of q. Thus if ϕ is the characteristic function of $|z_1|^2 - |z_2|^2$ where $z' = \{z_1, z_2\}$ is normal $(0, M)$ then

$$F(q) = \frac{1}{2\pi} \int_{-\infty}^{\infty} \phi^N(t) e^{-iqt} dt \quad . \tag{3.3}$$

If we let $r_1 = |z_1|$ and $r_2 = |z_2|$, then

$$\phi(t) = \mathscr{E} \exp [i(r_1^2 - r_2^2)t] = \int_{-\infty}^{\infty} \int_{-\infty}^{\infty} \exp [i(r_1^2 - r_2^2)t] g(r_1, r_2) dr_1 dr_2$$

where g is given by (2.2). Now let

$$M^{-1} = \begin{bmatrix} \mu & \lambda \\ \bar{\lambda} & \nu \end{bmatrix} \tag{3.4}$$

[see (2.3)]. Then expansion of the Bessel function in g and term by term integration leads to

$$\begin{aligned} \phi(t) &= \frac{1}{\det M} \frac{1}{(\mu - it)(\nu + it) - |\lambda|^2} \\ &= \frac{\det M^{-1}}{[\frac{1}{2}(B + A) - it][\frac{1}{2}(B - A) + it]} \end{aligned} \tag{3.5}$$

where

$$A = \mu - \nu = \operatorname{tr} JM^{-1}$$

$$B = [(\mu - \nu)^2 + 4(\mu\nu - |\lambda|^2)]^{1/2} = [(\operatorname{tr} JM^{-1})^2 + 4 \det M^{-1}]^{1/2} \quad .$$

Note that

$$B > |A| \quad . \tag{3.6}$$

For Re $a > 0$, Re $b > 0$, Re $\beta > \frac{1}{2}$ we have the Fourier transform formula

$$\frac{1}{2\pi} \int_{-\infty}^{\infty} (a + it)^{-\beta} (b - it)^{-\beta} \exp(-iqt) dt$$

$$= \frac{1}{\pi^{\frac{1}{2}}\Gamma(\beta)}\left(\frac{|q|}{a+b}\right)^{\beta-1/2} \exp\left[-\tfrac{1}{2}(b-a)q\right] K_{\beta-1/2}(\tfrac{1}{2}(a+b)|q|) \quad , \qquad (3.7)$$

$$-\infty < q < \infty \quad ,$$

where $K_{\beta-1/2}$ is the modified Bessel function of the second kind and order $\beta - \frac{1}{2}$. Thus from (3.3), (3.5), and (3.7),

$$F(q) = \frac{|q|^{N-1/2}}{\pi^{1/2}(N-1)!(\det \boldsymbol{M})^N B^{N-1/2}} e^{-Aq/2} K_{N-1/2}(\tfrac{1}{2} B|q|) \quad , \qquad (3.8)$$

$$-\infty < q < \infty \quad .$$

Under the hypothesis $\mathscr{H}_0$, $\boldsymbol{R} = \boldsymbol{R}_0$ and $\boldsymbol{M} = \boldsymbol{M}_0 = \boldsymbol{P}^{-1}\boldsymbol{R}_0\bar{\boldsymbol{P}}'^{-1}$. Since $\boldsymbol{J} = \boldsymbol{J}^{-1}$ we may write

$$A_0 = \operatorname{tr} \boldsymbol{J}\boldsymbol{M}_0^{-1} = \operatorname{tr} \boldsymbol{J}^{-1}\boldsymbol{M}_0^{-1} = \operatorname{tr} \boldsymbol{\Sigma}^{-1}\boldsymbol{R}_0^{-1}$$

$$= \frac{(-S_d) + S_0 - S_1}{(-S_d)} \qquad (3.9)$$

where S_0, S_1, S_d have been defined in (1.8). Note that since $\boldsymbol{\Sigma}$ is indefinite,

$$S_d = \det(\boldsymbol{R}_1 - \boldsymbol{R}_0) < 0 \quad ,$$

and that (3.9) is identical with (2.9). From (3.1)

$$|\det \boldsymbol{P}|^2(\det \boldsymbol{\Sigma}) = -1$$

and hence

$$\det \boldsymbol{M}_0 = |\det \boldsymbol{P}|^{-2}(\det \boldsymbol{R}_0) = \frac{(-S_d)}{S_1} \quad .$$

Thus

$$B_0 = [(\operatorname{tr} \boldsymbol{J}\boldsymbol{M}_0^{-1})^2 + 4 \det \boldsymbol{M}_0^{-1}]^{1/2}$$

$$= \frac{1}{(-S_d)}[((-S_d) + S_0 + S_1)^2 - 4S_0S_1]^{1/2} \quad .$$

Under the alternative hypothesis $\mathscr{H}_1$, $\boldsymbol{R} = \boldsymbol{R}_1$ and $\boldsymbol{M} = \boldsymbol{M}_1 = \boldsymbol{P}^{-1}\boldsymbol{R}_1\bar{\boldsymbol{P}}'^{-1}$. As above we show that

$$\det \boldsymbol{M}_1 = \frac{(-S_d)}{S_0}$$

$$A_1 = \operatorname{tr} \boldsymbol{J}\boldsymbol{M}_1^{-1}$$

$$= \frac{S_0 - S_1 - (-S_d)}{(-S_d)} \qquad (3.10)$$

[which is identical with (2.10)] and

$$B_1 = B_0 \quad .$$

We shall call their common value B.

To summarize, then, under the hypothesis $\mathscr{H}_0$, the frequency function of q is

$$F(q\,|\,\mathscr{H}_0) = \frac{S_1^N |q|^{N-1/2}}{\pi^{1/2}(N-1)!(-S_d)^N \, B^{N-1/2}} \exp\left(-\tfrac{1}{2}A_0 q\right) K_{N-1/2}\left(\tfrac{1}{2}B|q|\right) \quad ,$$

$$-\infty < q < \infty \quad , \qquad (3.11)$$

and under the alternative hypothesis $\mathscr{H}_1$, the frequency function of q is

$$F(q\,|\,\mathscr{H}_1) = \frac{S_0^N |q|^{N-1/2}}{\pi^{1/2}(N-1)!\,(-S_d)^N \, B^{N-1/2}} \exp\left(-\tfrac{1}{2}A_1 q\right) K_{N-1/2}\left(\tfrac{1}{2}B|q|\right) \quad ,$$

$$-\infty < q < \infty \quad , \qquad (3.12)$$

where A_0 and A_1 are given by (3.9) and (3.10) while

$$\begin{aligned} B &= \frac{1}{|S_d|} \left[((-S_d) + S_0 + S_1)^2 - 4S_0 S_1\right]^{1/2} \\ &= \frac{1}{|S_d|} \left(S_1^2 S_{01}^2 - 4S_0 S_1\right)^{1/2} \end{aligned} \qquad (3.13)$$

[which is identical with (2.13)].

4. The Semidefinite Case

Let $\boldsymbol{y}_n$, $1 \leqq n \leqq N$, be N independent two-dimensional complex Gaussian vectors with means zero and common positive definite covariance matrix $\boldsymbol{R}$. We wish to find the distribution of

$$q = \sum_{n=1}^{N} \bar{\boldsymbol{y}}_n' \boldsymbol{\Sigma} \boldsymbol{y}_n$$

under the hypothesis $\boldsymbol{R} = \boldsymbol{R}_0$ and under the alternative hypothesis $\boldsymbol{R} = \boldsymbol{R}_1$ where $\boldsymbol{\Sigma} = \boldsymbol{R}_0^{-1} - \boldsymbol{R}_1^{-1}$ is semidefinite.

A semidefinite matrix may be either positive semidefinite (that is, nonnegative definite) or negative semidefinite (that is, nonpositive definite). Using the same argument as in Section 2, we may assume, without loss of generality, that $\boldsymbol{\Sigma}$ is nonnegative definite. Thus there exists a nonsingular matrix $\boldsymbol{P}$ such that

$$\bar{\boldsymbol{P}}' \boldsymbol{\Sigma} \boldsymbol{P} = \boldsymbol{K} \qquad (4.1)$$

where

$$\boldsymbol{K} = \begin{bmatrix} 1 & 0 \\ 0 & 0 \end{bmatrix} \quad . \tag{4.2}$$

Let

$$\boldsymbol{z}_n = \boldsymbol{P}^{-1}\boldsymbol{y}_n \quad .$$

Then

$$\bar{\boldsymbol{y}}_n' \boldsymbol{\Sigma} \boldsymbol{y}_n = \bar{\boldsymbol{z}}_n' \bar{\boldsymbol{P}}' \boldsymbol{\Sigma} \boldsymbol{P} \boldsymbol{z}_n = |z_{n,1}|^2$$

where $\boldsymbol{z}_n' = \{z_{n,1}, z_{n,2}\}$, and

$$q = \sum_{n=1}^{N} |z_{n,1}|^2 \quad .$$

Also

$$\mathscr{E}\boldsymbol{z}_n = \boldsymbol{0}$$

and

$$\mathscr{E}\boldsymbol{z}_n \bar{\boldsymbol{z}}_n' = \boldsymbol{P}^{-1}\boldsymbol{R}\bar{\boldsymbol{P}}'^{-1} = \text{(say)} \quad \boldsymbol{M}$$

where $\boldsymbol{M}$ is positive definite.

As in the previous two sections we wish to compute the frequency function F of q. Thus if ϕ is the characteristic function of $|z_1|^2$ where $\boldsymbol{z}' = \{z_1, z_2\}$ is normal $(\boldsymbol{0}, \boldsymbol{M})$, then

$$F(q) = \frac{1}{2\pi} \int_{-\infty}^{\infty} \phi^N(t) e^{-iqt} dt \quad . \tag{4.3}$$

If, as in the previous two sections, we write

$$\boldsymbol{M}^{-1} = \begin{bmatrix} \mu & \lambda \\ \bar{\lambda} & \nu \end{bmatrix} \tag{4.4}$$

[see (2.3), (3.4)] and *not*, as in the previous two sections, let

$$A = \operatorname{tr} \boldsymbol{KM} = \nu(\det \boldsymbol{M}) > 0 \quad ,$$

then the probability density function g of $r_1 = |z_1|$ is [see (3.6) of Chapter III]

$$\begin{aligned} g(r_1) &= \frac{2r_1}{A} \exp\left(\frac{-r_1^2}{A}\right) \quad , & r_1 &\geqq 0 \quad , \\ &= 0, & r_1 &< 0 \quad . \end{aligned}$$

The corresponding characteristic function ϕ of r_1^2 is

$$\phi(t) = \mathscr{E} e^{ir_1^2 t} = \int_{-\infty}^{\infty} e^{ir_1^2 t} g(r_1) dr_1$$
$$= \frac{1}{1 - iAt} \quad .$$

For Re $a > 0$, Re $\beta > 0$ we have the inversion formula

$$\frac{1}{2\pi} \int_{-\infty}^{\infty} (a - it)^{-\beta} e^{-iqt} dt = \frac{q^{\beta-1} e^{-aq}}{\Gamma(\beta)} \quad , \qquad q > 0 \quad ,$$
$$= 0 \quad , \qquad q < 0 \quad ,$$

and thus

$$F(q) = \frac{q^{N-1}}{(N-1)! A^N} e^{-q/A} \quad , \qquad q > 0 \quad ,$$
$$= 0 \quad , \qquad q < 0 \quad . \tag{4.5}$$

Under the hypothesis $\mathscr{H}_0$, $\boldsymbol{R} = \boldsymbol{R}_0$ and $\boldsymbol{M} = \boldsymbol{M}_0 = \boldsymbol{P}^{-1} \boldsymbol{R}_0 \bar{\boldsymbol{P}}'^{-1}$. We shall write

$$A_0 = \operatorname{tr} \boldsymbol{K} \boldsymbol{M}_0 = \operatorname{tr} \boldsymbol{\Sigma} \boldsymbol{R}_0$$
$$= \operatorname{tr}(\boldsymbol{R}_0^{-1} - \boldsymbol{R}_1^{-1}) \boldsymbol{R}_0$$
$$= 2 - \operatorname{tr} \boldsymbol{R}_1^{-1} \boldsymbol{R}_0 = 2 - S_{01} \quad .$$

Since $\boldsymbol{\Sigma}$ is singular,

$$S_d = \det(\boldsymbol{R}_1 - \boldsymbol{R}_0) = 0$$

and [see (1.9)]

$$(\det \boldsymbol{R}_1)(\operatorname{tr} \boldsymbol{R}_1^{-1} \boldsymbol{R}_0) = \det \boldsymbol{R}_0 + \det \boldsymbol{R}_1 \quad .$$

Hence

$$A_0 = \frac{S_1 - S_0}{S_1} \quad . \tag{4.6}$$

Under the alternative hypothesis $\mathscr{H}_1$, $\boldsymbol{R} = \boldsymbol{R}_1$ and $\boldsymbol{M} = \boldsymbol{M}_1 = \boldsymbol{P}^{-1} \boldsymbol{R}_1 \bar{\boldsymbol{P}}'^{-1}$. As above we show that

$$A_1 = \frac{S_1 - S_0}{S_0} \quad . \tag{4.7}$$

To summarize, then, under the hypothesis $\mathscr{H}_0$, the frequency function of q is

$$
\begin{aligned}
F(q\,|\,\mathscr{H}_0) &= \frac{q^{N-1}}{(N-1)!\,A_0^N}\,e^{-q/A_0} \quad , \qquad q > 0 \quad , \\
&= 0 \quad , \qquad\qquad\qquad\qquad\quad q < 0 \quad ,
\end{aligned}
\tag{4.8}
$$

and under the alternative hypothesis $\mathscr{H}_1$, the frequency function of q is

$$
\begin{aligned}
F(q\,|\,\mathscr{H}_1) &= \frac{q^{N-1}}{(N-1)!\,A_1^N}\,e^{-q/A_1} \quad , \qquad q > 0 \quad , \\
&= 0 \quad , \qquad\qquad\qquad\qquad\quad q < 0 \quad ,
\end{aligned}
\tag{4.9}
$$

where A_0 and A_1 are given by (4.6) and (4.7).

5. Some Additional Comments

In certain applications, for example, the Gaussian approximations considered below, the moments of the likelihood ratio statistic q are of interest. Since we have explicit formulas for the density function F of q [(2.11) and (2.12) for the positive definite case, (3.11) and (3.12) for the indefinite case, (4.8) and (4.9) for the nonnegative definite case], we could compute the moments directly from the formula

$$
\mathscr{E}(q^\nu\,|\,\mathscr{H}_j) = \int_{-\infty}^{\infty} q^\nu F(q\,|\,\mathscr{H}_j)dq \quad , \qquad \nu = 0, 1, \cdots \qquad j = 0, 1 \quad .
$$

However, if we are interested only in the means and variances, it is perhaps easier to use the moment generating function (at least for the positive definite and indefinite cases).

Suppose, then, $m(\xi)$ is the moment generating function for $\bar{\boldsymbol{y}}_n' \boldsymbol{\Sigma} \boldsymbol{y}_n$. Then

$$
\mathscr{E}q = Nm'(0) \tag{5.1}
$$

and

$$
\text{Var}\, q = N\{m^{ii}(0) - [m'(0)]^2\} \quad . \tag{5.2}
$$

Now if we make the change of variable $t = -i\xi$ in the characteristic function ϕ of $\bar{\boldsymbol{y}}_n' \boldsymbol{\Sigma} \boldsymbol{y}_n$[see (2.4)], then for $|\xi|$ small,

$$
m(\xi) = \phi(-i\xi) = \frac{\det \boldsymbol{M}^{-1}}{[\frac{1}{2}(A+B) - \xi][\frac{1}{2}(A-B) - \xi]}
$$

is the moment generating function in the positive definite case. Thus, using (5.1) and (5.2), we find that

$$a_0 = \mathscr{E}(q|\mathscr{H}_0) = N\left(\frac{S_1 - S_0 + S_d}{S_1}\right) \tag{5.3}$$

$$\sigma_0^2 = \operatorname{Var}(q|\mathscr{H}_0) = N\left[\frac{(S_1 - S_0 + S_d)^2 - 2S_1S_d}{S_1^2}\right] \tag{5.4}$$

$$a_1 = \mathscr{E}(q|\mathscr{H}_1) = N\left(\frac{S_1 - S_0 - S_d}{S_0}\right) \tag{5.5}$$

$$\sigma_1^2 = \operatorname{Var}(q|\mathscr{H}_1) = N\left[\frac{(S_1 - S_0 - S_d)^2 - 2S_0S_d}{S_0^2}\right] \tag{5.6}$$

in the positive definite case, and from (3.5) we find that these formulas are also valid in the indefinite case. [One could also use Lemma 2.2 of Chapter V to derive the above formulas (see page 192)]. For the nonnegative definite case we obtain

$$a_0 = \mathscr{E}(q|\mathscr{H}_0) = N\left(\frac{S_1 - S_0}{S_1}\right)$$

$$\sigma_0^2 = \operatorname{Var}(q|\mathscr{H}_0) = N\left(\frac{S_1 - S_0}{S_1}\right)^2 = \frac{1}{N}\, a_0^2$$

$$a_1 = \mathscr{E}(q|\mathscr{H}_1) = N\left(\frac{S_1 - S_0}{S_0}\right)$$

$$\sigma_1^2 = \operatorname{Var}(q|\mathscr{H}_1) = N\left(\frac{S_1 - S_0}{S_0}\right)^2 = \frac{1}{N}\, a_1^2$$

directly from (4.8) and (4.9), which, we note are identical with (5.3)–(5.6) since in this case $S_d = 0$.

More generally, from (2.6) we obtain

$$\mathscr{E}q^\nu = \frac{\Gamma(2N+\nu)}{\Gamma(2N)(\det \boldsymbol{M})^N(\frac{1}{2}A)^{2N+\nu}} \times {}_2F_1\left(N + \tfrac{1}{2}\nu, N + \tfrac{1}{2}\nu + \tfrac{1}{2}, N + \tfrac{1}{2}; \frac{B^2}{A^2}\right) \quad , \qquad \operatorname{Re}\nu > -2N \quad ,$$

in the positive definite case, and from (3.8) we obtain

$$\mathscr{E}q^\nu = \frac{\Gamma(2N+\nu)\Gamma(\nu+1)2^{2N+\nu}}{\Gamma(N+\nu+1)\Gamma(N)(\det \boldsymbol{M})^N} \times \left[\frac{1}{(A+B)^{2N+\nu}}\,{}_2F_1\left(2N+\nu, N, N+\nu+1; \frac{A-B}{A+B}\right) + \frac{(-1)^\nu}{(B-A)^{2N+\nu}}\,{}_2F_1\left(2N+\nu, N, N+\nu+1; \frac{A+B}{A-B}\right)\right] \quad ,$$

$$\nu = 0, 1, \cdots \quad ,$$

in the indefinite case. In the formulas above ${}_2F_1$ is the hypergeometric function. Trivially, from (4.5)

$$\mathscr{E} q^\nu = \frac{\Gamma(N+\nu)}{\Gamma(N)} A^\nu \quad , \qquad \mathrm{Re}\, \nu > -N \quad ,$$

in the nonnegative definite case.

If N is a positive integer, the modified Bessel functions $I_{N-1/2}$ and $K_{N-1/2}$ are elementary functions, viz.:

$$I_{N-1/2}(\tfrac{1}{2}z) = \left(\frac{z}{\pi}\right)^{1/2} \sum_{j=1}^{N} \frac{(N+j-2)!}{(j-1)!\,(N-j)!} z^{-j}[(-1)^N e^{-z/2} - (-1)^j e^{z/2}]$$

and

$$K_{N-1/2}(\tfrac{1}{2}|z|) = (\pi|z|)^{1/2} e^{-|z|/2} \sum_{j=1}^{N} \frac{(N+j-2)!}{(j-1)!\,(N-j)!} |z|^{-j} \quad .$$

Thus in all three cases, the cumulative density function $\int_K^\infty F(q|\mathscr{H}_j)dq$, $j = 0,1$, may be evaluated without resort to numerical integration. With a given α and a fixed N, we may use an interative process (most conveniently carried out on a computer) to determine K from the equation

$$\alpha = \int_K^\infty F(q|\mathscr{H}_0)dq \quad .$$

Once the most powerful critical region has been determined, then the probability β,

$$\beta = 1 - \int_K^\infty F(q|\mathscr{H}_1)dq \quad ,$$

of committing a Type II error is minimum.

If the sample size N is large, it is sometimes convenient to use a Gaussian approximation. Thus q is approximately normal (a_j, σ_j^2) under the hypothesis $\mathscr{H}_j$, $j = 0, 1$ [see (5.3)–(5.6)].

Another problem of interest is to determine *a priori* (if possible) whether one is dealing with the definite or the indefinite case. In this connection we shall prove Theorem 5.1. Toward this end consider the fundamental matrix

$$\boldsymbol{\Sigma} = \boldsymbol{R}_0^{-1} - \boldsymbol{R}_1^{-1}$$

[see (1.5)] where

$$\boldsymbol{R}_0 = \begin{bmatrix} R_0(0) & R_0(h) \\ \bar{R}_0(h) & R_0(0) \end{bmatrix} \quad , \qquad h \neq 0 \quad , \tag{5.7}$$

and

$$\boldsymbol{R}_1 = \begin{bmatrix} R_1(0) & R_1(h) \\ \bar{R}_1(h) & R_1(0) \end{bmatrix} \quad , \qquad h \neq 0 \quad , \tag{5.8}$$

are positive definite. Then we shall prove:

Theorem 5.1. If $R_1(0) = R_0(0)$, then $\boldsymbol{\Sigma}$ is indefinite. If $\boldsymbol{\Sigma}$ is definite and if $R_1(0) < R_0(0)$, then $\boldsymbol{\Sigma}$ is negative definite. If $\boldsymbol{\Sigma}$ is definite and if $R_1(0) > R_0(0)$, then $\boldsymbol{\Sigma}$ is positive definite.

Proof. We recall [see (1.8)] that if

$$S_d = \det(\boldsymbol{R}_1 - \boldsymbol{R}_0) < 0 \quad ,$$

then $\boldsymbol{\Sigma}$ is indefinite. But

$$S_d = [R_1(0) - R_0(0)]^2 - |R_1(h) - R_0(h)|^2 \quad .$$

Hence if $R_1(0) = R_0(0)$, then $S_d < 0$ and $\boldsymbol{\Sigma}$ is indefinite.

Now suppose $\boldsymbol{\Sigma}$ is definite, that is, $S_d > 0$. Then if $\operatorname{tr} \boldsymbol{\Sigma} > 0$, the matrix $\boldsymbol{\Sigma}$ is positive definite, and conversely; and if $\operatorname{tr} \boldsymbol{\Sigma} < 0$, the matrix $\boldsymbol{\Sigma}$ is negative definite, and conversely. But $\operatorname{tr} \boldsymbol{\Sigma} > 0$ implies $A_0 = \operatorname{tr} \boldsymbol{\Sigma}^{-1} \boldsymbol{R}_0^{-1} > 0$ since $\boldsymbol{R}_0$ is positive definite (and $A_1 > 0$) and conversely. Similarly, $\operatorname{tr} \boldsymbol{\Sigma} < 0$ implies $A_0 < 0$ (and $A_1 < 0$) and conversely.

We shall now show that if $S_d > 0$ and $R_1(0) < R_0(0)$, then $A_0 < 0$ and hence $\boldsymbol{\Sigma}$ is negative definite. Since $S_d > 0$, to show that $A_0 < 0$ is equivalent to proving that $S_d A_0 < 0$. But from (2.9)

$$S_d A_0 = S_1 - S_0 + S_d \quad .$$

Thus from (1.8), (5.7), and (5.8),

$$\begin{aligned} S_d A_0 &= S_1 - S_0 + S_d \\ &= R_1^2(0) - |R_1(h)|^2 - R_0^2(0) + |R_0(h)|^2 + [R_1(0) - R_0(0)]^2 \\ &\quad - |R_1(h) - R_0(h)|^2 \\ &= 2R_1(0)\,[R_1(0) - R_0(0)] - |R_1(h)|^2 - |R_1(h) - R_0(h)|^2 \\ &\quad + |R_0(h)|^2 \quad . \end{aligned} \tag{5.9}$$

Now if we assume $R_1(0) < R_0(0)$ and use the facts that S_d and S_0 are positive, we obtain

$$2R_1(0)\,[R_1(0) - R_0(0)] < -\,2|R_1(h)|\,|R_1(h) - R_0(h)| \quad .$$

Substituting this result in (5.9) then yields

$$
\begin{aligned}
S_d A_0 &< -[|R_1(h)| + |R_1(h) - R_0(h)|]^2 + |R_0(h)|^2 \\
&= [|R_0(h)| + |R_1(h)| + |R_1(h) - R_0(h)|] \\
&\quad \times [|R_0(h)| - |R_1(h)| - |R_1(h) - R_0(h)|] \quad .
\end{aligned}
$$

But

$$
|R_0(h)| - |R_1(h)| - |R_1(h) - R_0(h)| \leqq 0 \quad .
$$

Thus $A_0 < 0$.

Now suppose $S_d > 0$ and $R_1(0) > R_0(0)$. Then

$$
\begin{aligned}
S_d A_0 &> 2|R_1(h)|\,|R_1(h) - R_0(h)| - |R_1(h)|^2 - |R_1(h) - R_0(h)|^2 + |R_0(h)|^2 \\
&= -[|R_1(h)| - |R_1(h) - R_0(h)|]^2 + |R_0(h)|^2 \\
&= [|R_0(h)| - |R_1(h)| + |R_1(h) - R_0(h)|] \\
&\quad \times [|R_0(h)| + |R_1(h)| - |R_1(h) - R_0(h)|] \quad .
\end{aligned}
$$

Both terms in brackets in the last inequality are nonnegative. Thus $A_0 > 0$ and the proof of the theorem is complete.

VIII

COMPLEX LINEAR LEAST SQUARES

0. Introduction

If a finite sequence of observations $z_1, \cdots, z_n$ is made at the times $t_1, \cdots, t_n$, respectively, one is invited to approximate the data by a "smooth" curve, say $z = \phi(t)$. In almost all practical situations the curve will not pass exactly through all the points. Hence one attempts to draw the "best" curve. "Best" is frequently interpreted in the mean square sense. That is, one tries to minimize the sum of the squares of the differences between the observed values z_k and $\phi(t_k)$. This is essentially the idea behind the theory of least squares. (Note that in such an analysis the notion of random variable need not enter into the discussion.) We shall develop certain portions of the theory of linear least squares from the point of view of complex stochastic processes. Our results parallel the real case.

To fix our ideas we illustrate the basic problem with a simple example (for the real case) in Section 1. In later sections we generalize the problem to higher dimensions and to the use of complex quantities. Before doing this, however, we find it convenient to make a slight digression in order to introduce the concept of differentiation of a scalar with respect to a complex matrix. This is a useful technique which we shall have occasion to use more than once. In Section 3 we derive the *normal equation* satisfied by the *least squares estimate* (LSE). After a discussion of various equivalent least squares models, we introduce the *best linear unbiased estimate* (BLUE) and the *Markov estimate* (ME). We next prove the Gauss–Markov theorem which expresses the relationship between the LSE and ME (for a certain class of covariance functions), and then give a necessary and sufficient condition that the LSE and ME be identical.

The efficiency of the LSE is defined in Section 6. We use the Kantorovich inequality to obtain a lower bound for the efficiency of a certain class of models. Next we turn to the problem of estimating the mean vector and the variance of a sample from a normal population. This leads to the *maximum likelihood estimate* (MLE) and the fact that the MLE of the mean is identical with the LSE. We also show that the maximum likelihood estimators form a set of sufficient statistics, and then explicitly determine their probability density functions. Finally, in Section 8, we briefly discuss regression functions.

1. The Basic Idea

Anyone who has taken a laboratory course in the physical sciences has faced the problem of attempting to interpret empirical data. For example, at a particular value of an independent variable, say x_t, one might observe a physical quantity with numerical value, say z_t. If a finite number, say T, of these observations is made, one may represent the data graphically as in Figure 1. A typical problem might be to find the "best" straight line that fits the data (see Figure 2). By "best" straight line one generally means a line with equation $z = \beta_1 x + \beta_2$ where β_1 and β_2 are so chosen that the sum

$$v = \sum_{t=1}^{T} [z_t - (\beta_1 x_t + \beta_2)]^2 \tag{1.1}$$

is minimized. If one computes $\partial v/\partial \beta_1$ and $\partial v/\partial \beta_2$ and sets these derivatives equal to zero, there results

$$\boldsymbol{\beta} = (\boldsymbol{H}'\boldsymbol{H})^{-1}\boldsymbol{H}'\boldsymbol{z} \tag{1.2}$$

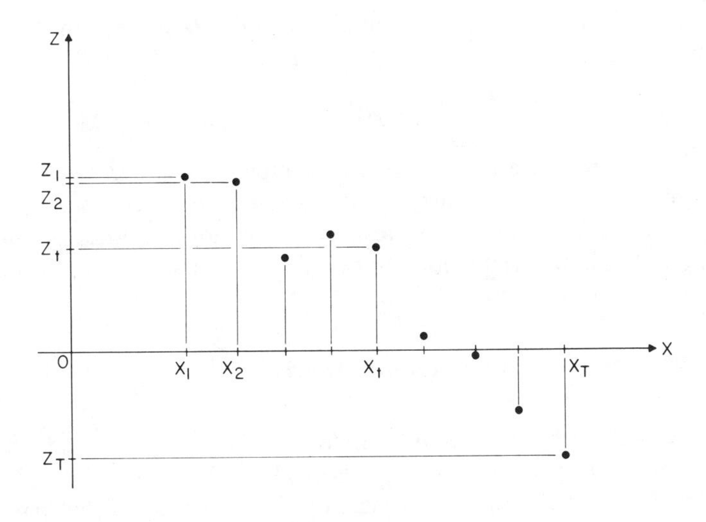

Fig. 1

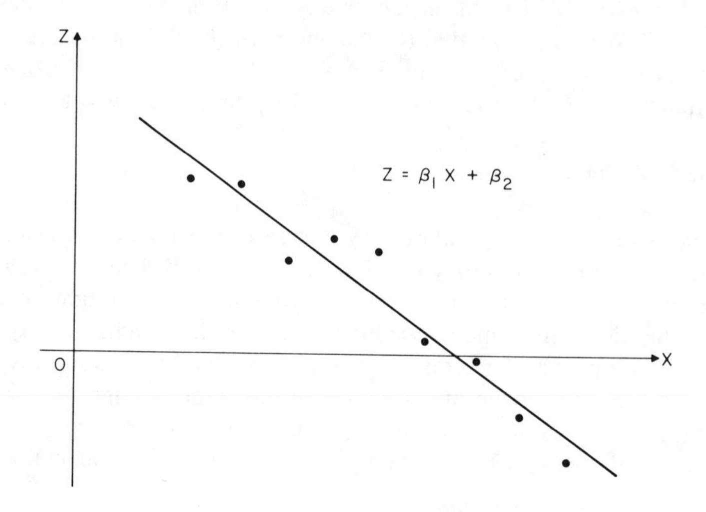

Fig. 2

where $\boldsymbol{\beta}' = \{\beta_1, \beta_2\}$, $\boldsymbol{z}' = \{z_1, \cdots, z_T\}$ and

$$\boldsymbol{H}' = \begin{bmatrix} x_1 & x_2 & \cdots & x_T \\ 1 & 1 & \cdots & 1 \end{bmatrix} . \tag{1.3}$$

If not all the x_t, $1 \leqq t \leqq T$, are identical, then det $\boldsymbol{H}'\boldsymbol{H} > 0$. In the matrix notation just introduced, (1.1) may be written as

$$v = |\boldsymbol{z} - \boldsymbol{H}\boldsymbol{\beta}|^2 . \tag{1.4}$$

We wish to generalize this problem to complex quantities and to vectors $\boldsymbol{\beta}$ of arbitrary dimension. To carry out this program most efficiently we need the concept of differentiation with respect to a vector. We therefore make a slight digression in the next section to explain this technique, and then return to our main theme in Section 3.

2. Differentiation with Respect to a Matrix

Let $\boldsymbol{Z} = \| Z_{jk} \|$ be an $n \times m$ complex matrix. Let g be a scalar-valued function of $\boldsymbol{Z}$. If we write $Z_{jk} = X_{jk} + iY_{jk}$ where $X_{jk} =$ Re Z_{jk} and $Y_{jk} =$ Im Z_{jk}, then g may be considered to be a function of the $2nm$ variables X_{jk}, Y_{jk}, $1 \leqq j \leqq n$, $1 \leqq k \leqq m$. Now suppose g is a differentiable function of these $2nm$ variables in some region of $2nm$-dimensional real Euclidean space. Then we define the derivative of g with respect to $\boldsymbol{Z}$ as the $n \times m$ matrix

$$\frac{dg}{d\boldsymbol{Z}} = \frac{1}{2} \left\| \frac{\partial g}{\partial X_{jk}} - i \frac{\partial g}{\partial Y_{jk}} \right\| \tag{2.1}$$

where the X_{jk}, Y_{jk} are to be thought of as $2nm$ independent variables.

For example, if $\boldsymbol{\xi}$ and $\boldsymbol{c}$ are n-dimensional complex vectors, then a direct application of this definition yields

$$\frac{d}{d\boldsymbol{\xi}}(\boldsymbol{c}'\boldsymbol{\xi}) = \boldsymbol{c}$$

and

$$\frac{d}{d\boldsymbol{\xi}}(\boldsymbol{c}'\bar{\boldsymbol{\xi}}) = \boldsymbol{0} \quad .$$

If $\boldsymbol{Q}$ is an arbitrary $n \times n$ complex matrix, then

$$\frac{d}{d\boldsymbol{\xi}}(\boldsymbol{\xi}'\boldsymbol{Q}\boldsymbol{\xi}) = (\boldsymbol{Q} + \boldsymbol{Q}')\,\boldsymbol{\xi}$$

$$\frac{d}{d\boldsymbol{\xi}}(\bar{\boldsymbol{\xi}}'\boldsymbol{Q}\boldsymbol{\xi}) = \boldsymbol{Q}'\bar{\boldsymbol{\xi}}$$

$$\frac{d}{d\boldsymbol{\xi}}(\boldsymbol{\xi}'\boldsymbol{Q}\bar{\boldsymbol{\xi}}) = \boldsymbol{Q}\bar{\boldsymbol{\xi}}$$

and

$$\frac{d}{d\boldsymbol{Q}} \det \boldsymbol{Q} = (\det \boldsymbol{Q})\, \boldsymbol{Q}'^{-1} \quad .$$

Furthermore, if $\boldsymbol{K}$ is an $m \times n$ complex matrix, $\boldsymbol{A}$ an $n \times m$ complex matrix, and $\boldsymbol{d}$ an m-dimensional complex vector, then

$$\frac{d}{d\boldsymbol{K}}(\boldsymbol{d}'\boldsymbol{K}\boldsymbol{c}) = \boldsymbol{d}\boldsymbol{c}'$$

$$\frac{d}{d\boldsymbol{K}} \operatorname{tr} \boldsymbol{A}\boldsymbol{K} = \boldsymbol{A}'$$

and if $\boldsymbol{\gamma}$ is also an n-dimensional complex vector,

$$\frac{d}{d\boldsymbol{K}}(\boldsymbol{c}'\boldsymbol{K}'\bar{\boldsymbol{K}}\boldsymbol{\gamma}) = \bar{\boldsymbol{K}}\boldsymbol{\gamma}\boldsymbol{c}' \quad .$$

Suppose now that g is real. Then [see (2.1)], the matrix equation

$$\frac{dg}{d\boldsymbol{Z}} = \boldsymbol{0} \tag{2.2}$$

is equivalent to the $2nm$ real scalar equations

$$\frac{\partial g}{\partial X_{jk}} = 0 \ , \qquad \frac{\partial g}{\partial Y_{jk}} = 0 \ , \qquad 1 \leqq j \leqq n \ , \qquad 1 \leqq k \leqq m \ . \tag{2.3}$$

Thus (2.2) represents a necessary condition that g have a stationary point in the sense of the calculus.

For example, let it be required to find an n-dimensional complex vector $\boldsymbol{\xi}$ which minimizes

$$g(\boldsymbol{\xi}) = |\boldsymbol{d} - \boldsymbol{K\xi}|^2 \tag{2.4}$$

(where $\boldsymbol{K}$ and $\boldsymbol{d}$ are as defined above). Then

$$\frac{dg}{d\boldsymbol{\xi}} = -\boldsymbol{K}'(\bar{\boldsymbol{d}} - \bar{\boldsymbol{K}}\bar{\boldsymbol{\xi}}) \ .$$

Setting this derivative equal to the n-dimensional zero vector and taking conjugates leads to

$$\bar{\boldsymbol{K}}'\boldsymbol{K\xi} = \bar{\boldsymbol{K}}'\boldsymbol{d} \ . \tag{2.5}$$

Thus if $\bar{\boldsymbol{K}}'\boldsymbol{K}$ is nonsingular,

$$\boldsymbol{\xi} = (\bar{\boldsymbol{K}}'\boldsymbol{K})^{-1}\bar{\boldsymbol{K}}'\boldsymbol{d} \ . \tag{2.6}$$

Of course, in this example, it is evident from an inspection of (2.4) that if the equation $\boldsymbol{d} - \boldsymbol{K\xi} = \boldsymbol{0}$ admits a solution $\boldsymbol{\xi}$, then $\boldsymbol{K\xi} = \boldsymbol{d}$ is the desired solution. But this equation (when multiplied by $\bar{\boldsymbol{K}}'$) is (2.5).

3. The Normal Equation

We now consider a generalization of the regression problem discussed in Section 1. Let $\boldsymbol{\beta}' = \{\beta_1, \cdots, \beta_p\}$ be a p-dimensional (complex) vector parameter. Let $\boldsymbol{H}$ be a $T \times p$ matrix of rank p. The elements of $\boldsymbol{H}$ are known, fixed, complex numbers. By analogy with (1.3) we call these elements *independent variables.* Let $\boldsymbol{z}' = \{z_1, \cdots, z_T\}$ be the T-dimensional (complex) vector of observables. By analogy with (1.1) we call the z_t *dependent variables.* Let $\boldsymbol{b}^*$ be a p-dimensional (complex) vector with the property that when $\boldsymbol{\beta}$ is replaced by $\boldsymbol{b}^*$ in $|\boldsymbol{z} - \boldsymbol{H\beta}|^2$, this expression is minimized. We call $\boldsymbol{b}^*$ the *least squares estimate* (LSE) of $\boldsymbol{\beta}$. In expanded notation, $\boldsymbol{b}^{*\prime} = \{b_1^*, \cdots, b_p^*\}$ is that value of $\boldsymbol{\beta}$ which minimizes

$$v = \sum_{t=1}^{T} \left| z_t - \sum_{j=1}^{p} H_{tj}\beta_j \right|^2 \tag{3.1}$$

(where $\boldsymbol{H} = \| H_{tj} \|$).

It is not difficult to find $\boldsymbol{b}^*$. We first write (3.1) in matrix form as

$$v = |\boldsymbol{z} - \boldsymbol{H\beta}|^2 \quad .$$

Then as in the previous section we deduce that

$$\boldsymbol{b}^* = (\bar{\boldsymbol{H}}'\boldsymbol{H})^{-1}\bar{\boldsymbol{H}}'\boldsymbol{z} \tag{3.2}$$

is the LSE of $\boldsymbol{\beta}$ [cf. (1.2) and (2.6)].

The matrix $\bar{\boldsymbol{H}}'\boldsymbol{H}$ in (3.2) is nonsingular. This follows from the more general result that if the rank of $\boldsymbol{H}$ is p and $\boldsymbol{\Sigma}$ is a $T \times T$ positive definite matrix, then $\bar{\boldsymbol{H}}'\boldsymbol{\Sigma}\boldsymbol{H}$ is positive definite (see Section 2 of Chapter II).

If we let

$$\boldsymbol{A} = \bar{\boldsymbol{H}}'\boldsymbol{H}$$

and

$$\boldsymbol{c} = \bar{\boldsymbol{H}}'\boldsymbol{z} \quad ,$$

then (3.2) may be written

$$\boldsymbol{b}^* = \boldsymbol{A}^{-1}\boldsymbol{c} \tag{3.3}$$

(and $\boldsymbol{Ab}^* = \boldsymbol{c}$). We call (3.3) the *normal equation.*

4. The Gauss–Markov Theorem

Suppose, as above, $\boldsymbol{z}' = \{z_1, \cdots, z_T\}$ is a T-dimensional observation vector, $\boldsymbol{\beta}' = \{\beta_1, \cdots, \beta_p\}$ is a p-dimensional parameter vector, and $\boldsymbol{H} = \| H_{tj} \|$ is a $T \times p$ constant matrix of rank p. Let

$$\begin{aligned} \mathscr{E}\boldsymbol{z} &= \boldsymbol{H\beta} \\ \mathscr{E}(\boldsymbol{z} - \boldsymbol{H\beta})(\bar{\boldsymbol{z}} - \bar{\boldsymbol{H}}\bar{\boldsymbol{\beta}})' &= \boldsymbol{\Sigma} \end{aligned} \tag{4.1}$$

where $\boldsymbol{\Sigma}$ is a $T \times T$ positive definite Hermitian covariance matrix. We shall refer to such a model as $\mathscr{M}_{Tp}[\boldsymbol{z}, \boldsymbol{\beta}, \boldsymbol{H}, \boldsymbol{\Sigma}]$.

Another way of describing $\mathscr{M}_{Tp}[\boldsymbol{z}, \boldsymbol{\beta}, \boldsymbol{H}, \boldsymbol{\Sigma}]$ is as follows: Let $\{z_t | t = \cdots, -1, 0, 1, \cdots\}$ be a complex stochastic process with discrete parameter t. Let $z_1, \cdots, z_T$ be observations from $\{z_t | t = \cdots, -1, 0, 1, \cdots \}$. Let the vector of observations $\boldsymbol{z}' = \{z_1, \cdots, z_T\}$ have a mean value $\boldsymbol{H\beta}$ and positive definite covariance matrix $\boldsymbol{\Sigma}$.

Still another description of $\mathscr{M}_{Tp}[\boldsymbol{z}, \boldsymbol{\beta}, \boldsymbol{H}, \boldsymbol{\Sigma}]$ is furnished by a consideration of

$$z_t = \sum_{j=1}^{p} H_{tj}\beta_j + u_t \quad , \qquad 1 \leqq t \leqq T \quad , \tag{4.2}$$

where the H_{tj}, $1 \leqq t \leqq T$, $1 \leqq j \leqq p$, are constants, the β_j, $1 \leqq j \leqq p$, are parameters to be estimated, and the u_t, $1 \leqq t \leqq T$, are random variables with means zero and positive definite covariance matrix $\boldsymbol{\Sigma}$. One may write (4.2) in vector notation as

$$\boldsymbol{z} = \boldsymbol{H\beta} + \boldsymbol{u} \tag{4.3}$$

where $\boldsymbol{u}' = \{u_1, \cdots, u_T\}$ and $\mathscr{E}\boldsymbol{u} = \boldsymbol{0}$, $\mathscr{E}\boldsymbol{u}\bar{\boldsymbol{u}}' = \boldsymbol{\Sigma}$. We call $\boldsymbol{H\beta}$ the *regression function* of $\boldsymbol{z}$ on $\boldsymbol{\beta}$ and call $\boldsymbol{u}$ the *residual.*

The fundamental problem is to find the "best" estimate of the parameter $\boldsymbol{\beta}$ based on the observables $z_1, \cdots, z_T$. To be more precise, let $\boldsymbol{\gamma}$ be an arbitrary but fixed p-dimensional vector. We shall find a T-dimensional vector $\boldsymbol{g}$ such that $\boldsymbol{g}'\boldsymbol{z}$ is a minimum variance, unbiased estimate of $\boldsymbol{\gamma}'\boldsymbol{\beta}$. Since $\boldsymbol{g}'\boldsymbol{z}$ is a linear combination of the observations $z_1, \cdots, z_T$, we see that $\boldsymbol{g}'\boldsymbol{z}$ will be a minimum variance, *linear,* unbiased estimate. We call such an estimate a *best linear unbiased estimate* (BLUE).

Let us find $\boldsymbol{g}$. If $\boldsymbol{g}'\boldsymbol{z}$ is to be unbiased we must have

$$\mathscr{E}\boldsymbol{g}'\boldsymbol{z} = \boldsymbol{g}'\boldsymbol{H\beta} = \boldsymbol{\gamma}'\boldsymbol{\beta}$$

which implies that

$$\boldsymbol{g}'\boldsymbol{H} = \boldsymbol{\gamma}' \quad . \tag{4.4}$$

Hence, for any p-dimensional vector $\boldsymbol{\lambda}$,

$$\boldsymbol{g}'\boldsymbol{H}\bar{\boldsymbol{\lambda}} + \bar{\boldsymbol{g}}'\bar{\boldsymbol{H}}\boldsymbol{\lambda} = \boldsymbol{\gamma}'\bar{\boldsymbol{\lambda}} + \bar{\boldsymbol{\gamma}}'\boldsymbol{\lambda}$$

is a real scalar. Since the variance of $\boldsymbol{g}'\boldsymbol{z}$ is

$$\operatorname{Var} \boldsymbol{g}'\boldsymbol{z} = \mathscr{E}(\boldsymbol{g}'\boldsymbol{z} - \boldsymbol{g}'\boldsymbol{H\beta})(\bar{\boldsymbol{g}}'\bar{\boldsymbol{z}} - \bar{\boldsymbol{g}}'\bar{\boldsymbol{H}}\bar{\boldsymbol{\beta}})' = \boldsymbol{g}'\boldsymbol{\Sigma}\bar{\boldsymbol{g}} \quad ,$$

we must minimize

$$v = \boldsymbol{g}'\boldsymbol{\Sigma}\bar{\boldsymbol{g}} - (\boldsymbol{g}'\boldsymbol{H}\bar{\boldsymbol{\lambda}} + \bar{\boldsymbol{g}}'\bar{\boldsymbol{H}}\boldsymbol{\lambda})$$

where $\boldsymbol{\lambda}$ plays the role of a Lagrange multiplier. That is, we have a constrained minimization problem. Setting the derivative

$$\frac{dv}{d\boldsymbol{g}} = \boldsymbol{\Sigma}\bar{\boldsymbol{g}} - \boldsymbol{H}\bar{\boldsymbol{\lambda}}$$

equal to the T-dimensional zero vector leads to

$$\boldsymbol{g}' = \boldsymbol{\lambda}'\bar{\boldsymbol{H}}'\boldsymbol{\Sigma}^{-1} \tag{4.5}$$

and by (4.4)

$$\boldsymbol{\gamma}' = \boldsymbol{g}'\boldsymbol{H} = \boldsymbol{\lambda}'(\bar{\boldsymbol{H}}'\boldsymbol{\Sigma}^{-1}\boldsymbol{H})$$

or, since $\bar{\boldsymbol{H}}'\boldsymbol{\Sigma}^{-1}\boldsymbol{H}$ is nonsingular (see Section 2 of Chapter II),

$$\boldsymbol{\lambda}' = \boldsymbol{\gamma}'(\bar{\boldsymbol{H}}'\boldsymbol{\Sigma}^{-1}\boldsymbol{H})^{-1} \quad .$$

Thus (4.5) yields

$$\boldsymbol{g}' = \boldsymbol{\gamma}'(\bar{\boldsymbol{H}}'\boldsymbol{\Sigma}^{-1}\boldsymbol{H})^{-1}\bar{\boldsymbol{H}}'\boldsymbol{\Sigma}^{-1}$$

and

$$\boldsymbol{g}'\boldsymbol{z} = \boldsymbol{\gamma}'(\bar{\boldsymbol{H}}'\boldsymbol{\Sigma}^{-1}\boldsymbol{H})^{-1}\bar{\boldsymbol{H}}'\boldsymbol{\Sigma}^{-1}\boldsymbol{z} \tag{4.6}$$

is the BLUE of $\boldsymbol{\gamma}'\boldsymbol{\beta}$.

Let $\boldsymbol{b}' = \{b_1, \cdots, b_p\}$ be the vector with the property that b_j, $1 \leqq j \leqq p$, is the BLUE of β_j. Then we call $\boldsymbol{b}$ the *Markov estimate* (ME) of $\boldsymbol{\beta}$. Now suppose, in particular, that $\boldsymbol{\gamma} = \boldsymbol{\delta}_k$ where $\boldsymbol{\delta}_k$ is a p-dimensional vector with one in the kth position and zeros elsewhere. That is, $\boldsymbol{\delta}_k' = \{\delta_{1k}, \cdots, \delta_{pk}\}$ where δ_{jk} is the Kronecker delta. Then $\beta_k = \boldsymbol{\delta}_k'\boldsymbol{\beta}$ and from (4.6), the BLUE of $\boldsymbol{\delta}_k'\boldsymbol{\beta}$ (= the kth component of $\boldsymbol{\beta}$) is $\boldsymbol{\delta}_k'(\bar{\boldsymbol{H}}'\boldsymbol{\Sigma}^{-1}\boldsymbol{H})^{-1}\bar{\boldsymbol{H}}'\boldsymbol{\Sigma}^{-1}\boldsymbol{z}$(= the kth component of $(\bar{\boldsymbol{H}}'\boldsymbol{\Sigma}^{-1}\boldsymbol{H})^{-1}\bar{\boldsymbol{H}}'\boldsymbol{\Sigma}^{-1}\boldsymbol{z}$.) Thus

$$\boldsymbol{b} = (\bar{\boldsymbol{H}}'\boldsymbol{\Sigma}^{-1}\boldsymbol{H})^{-1}\bar{\boldsymbol{H}}'\boldsymbol{\Sigma}^{-1}\boldsymbol{z} \tag{4.7}$$

is the ME of $\boldsymbol{\beta}$. We also see that the BLUE of $\boldsymbol{\gamma}'\boldsymbol{\beta}$ is $\boldsymbol{\gamma}'\boldsymbol{b}$ where $\boldsymbol{b}$ is the ME of $\boldsymbol{\beta}$ and $\boldsymbol{\gamma}$ is an arbitrary, known, p-dimensional vector.

The LSE of $\boldsymbol{\beta}$ [see (3.2)],

$$\boldsymbol{b}^* = (\bar{\boldsymbol{H}}'\boldsymbol{H})^{-1}\bar{\boldsymbol{H}}'\boldsymbol{z} \tag{4.8}$$

is also a linear unbiased estimate of $\boldsymbol{\beta}$. [The form of (4.8) demonstrates its linearity in $\boldsymbol{z}$, and $\mathscr{E}\boldsymbol{b}^* = (\bar{\boldsymbol{H}}'\boldsymbol{H})^{-1}\bar{\boldsymbol{H}}'\mathscr{E}\boldsymbol{z} = (\bar{\boldsymbol{H}}'\boldsymbol{H})^{-1}\bar{\boldsymbol{H}}'\boldsymbol{H}\boldsymbol{\beta} = \boldsymbol{\beta}$.] The covariance of $\boldsymbol{b}^*$ is

$$\mathscr{E}(\boldsymbol{b}^* - \boldsymbol{\beta})(\bar{\boldsymbol{b}}^* - \bar{\boldsymbol{\beta}})' = (\bar{\boldsymbol{H}}'\boldsymbol{H})^{-1}\bar{\boldsymbol{H}}'\boldsymbol{\Sigma}\boldsymbol{H}(\bar{\boldsymbol{H}}'\boldsymbol{H})^{-1} \quad . \tag{4.9}$$

The covariance matrix of the ME of $\boldsymbol{\beta}$ is

$$\mathscr{E}(\boldsymbol{b} - \boldsymbol{\beta})(\bar{\boldsymbol{b}} - \bar{\boldsymbol{\beta}})' = (\bar{\boldsymbol{H}}'\boldsymbol{\Sigma}^{-1}\boldsymbol{H})^{-1} \quad . \tag{4.10}$$

Since $\boldsymbol{\gamma}'\boldsymbol{b}$ is the BLUE of $\boldsymbol{\gamma}'\boldsymbol{\beta}$ for arbitrary $\boldsymbol{\gamma}$, we infer that

$$\text{Var}\,\boldsymbol{\gamma}'\boldsymbol{b} \leqq \text{Var}\,\boldsymbol{\gamma}'\boldsymbol{b}^*$$

or, in expanded notation,

$$\boldsymbol{\gamma}'[(\bar{\boldsymbol{H}}'\boldsymbol{H})^{-1}\bar{\boldsymbol{H}}'\boldsymbol{\Sigma}\boldsymbol{H}(\bar{\boldsymbol{H}}'\boldsymbol{H})^{-1} - (\bar{\boldsymbol{H}}'\boldsymbol{\Sigma}^{-1}\boldsymbol{H})^{-1}]\,\bar{\boldsymbol{\gamma}} \geqq 0 \quad .$$

But this in turn is equivalent to saying that the matrix

$$(\bar{\boldsymbol{H}}'\boldsymbol{H})^{-1}(\bar{\boldsymbol{H}}'\boldsymbol{\Sigma}\boldsymbol{H})(\bar{\boldsymbol{H}}'\boldsymbol{H})^{-1} - (\bar{\boldsymbol{H}}'\boldsymbol{\Sigma}^{-1}\boldsymbol{H})^{-1} \tag{4.11}$$

is nonnegative definite.

If in particular $\boldsymbol{\Sigma} = \sigma^2\boldsymbol{I}$, where $\boldsymbol{I}$ is the $T \times T$ identity matrix, then [cf. (4.7) and (4.8)] $\boldsymbol{b} = \boldsymbol{b}^*$. That is, the LSE and ME are identical. This last sentence is a statement of the *Gauss–Markov theorem*. Formally stated:

Theorem 4.1 (Gauss–Markov Theorem). Consider the model $\mathscr{M}_{Tp}[\boldsymbol{z}, \boldsymbol{\beta}, \boldsymbol{H}, \sigma^2\boldsymbol{I}]$. Let $\boldsymbol{b}^*$ be the least squares estimate of $\boldsymbol{\beta}$. Then the components of $\boldsymbol{b}^*$ are the minimum variance linear unbiased estimates (among all linear unbiased estimates) of the corresponding components of $\boldsymbol{\beta}$.

Corollary 4.1. Consider the model $\mathscr{M}_{Tp}[\boldsymbol{z}, \boldsymbol{\beta}, \boldsymbol{H}, \sigma^2\boldsymbol{I}]$. Let $\boldsymbol{b}^*$ be the least squares estimate of $\boldsymbol{\beta}$. Let $\boldsymbol{\gamma}$ be an arbitrary, but fixed p-dimensional vector. Then the minimum variance linear unbiased estimate of $\boldsymbol{\gamma}'\boldsymbol{\beta}$ is $\boldsymbol{\gamma}'\boldsymbol{b}^*$.

Trivially we note that when $\boldsymbol{\Sigma} = \sigma^2\boldsymbol{I}$, the covariance matrix of $\boldsymbol{b}$ $(= \boldsymbol{b}^*)$ becomes

$$\sigma^2(\bar{\boldsymbol{H}}'\boldsymbol{H})^{-1} \quad . \tag{4.12}$$

5. Equivalence of Least Squares and Markov Estimates

If in the model $\mathscr{M}_{Tp}[\boldsymbol{z}, \boldsymbol{\beta}, \boldsymbol{H}, \boldsymbol{\Sigma}]$ the covariance matrix $\boldsymbol{\Sigma}$ is a scalar multiple of the identity matrix, $\boldsymbol{\Sigma} = \sigma^2\boldsymbol{I}$, then the ME $\boldsymbol{b}$ and LSE $\boldsymbol{b}^*$ are identical. In this section we wish to establish necessary and sufficient conditions under which $\boldsymbol{b}$ equals $\boldsymbol{b}^*$. The results are given in Theorem 5.1. However, before proving this theorem, we need the following lemma from matrix theory (which is also used in the next section):

Lemma 5.1. Let $\boldsymbol{G}$ be a positive definite $n \times n$ Hermitian matrix and let $\boldsymbol{\Gamma}$ be a nonnegative definite $n \times n$ Hermitian matrix. Then there exists a nonsingular $n \times n$ matrix $\boldsymbol{Q}$ such that

$$\bar{\boldsymbol{Q}}'\boldsymbol{G}\boldsymbol{Q} = \boldsymbol{I} \tag{5.1}$$

and

$$\bar{\boldsymbol{Q}}'\boldsymbol{\Gamma}\boldsymbol{Q} = \boldsymbol{M} \tag{5.2}$$

where $\boldsymbol{I}$ is the $n \times n$ identity matrix and $\boldsymbol{M} = \| \mu_j \delta_{jk} \|$ is a diagonal matrix where the μ_j, $1 \leqq j \leqq n$, are the roots of the characteristic equation det $(\boldsymbol{\Gamma} - \mu\boldsymbol{G}) = 0$. Furthermore, if $\boldsymbol{G} - \boldsymbol{\Gamma}$ is nonnegative definite, then

$$\det \boldsymbol{G} - \det \boldsymbol{\Gamma} \geqq 0 \quad . \tag{5.3}$$

Proof. Let $\boldsymbol{P}$ be a nonsingular matrix such that

$$\bar{\boldsymbol{P}}'\boldsymbol{G}\boldsymbol{P} = \boldsymbol{I} \quad . \tag{5.4}$$

Since $\bar{\boldsymbol{P}}'\boldsymbol{\Gamma}\boldsymbol{P}$ is Hermitian, there exists a unitary matrix $\boldsymbol{U}$ such that

$$\bar{\boldsymbol{U}}'(\bar{\boldsymbol{P}}'\boldsymbol{\Gamma}\boldsymbol{P})\,\boldsymbol{U} = \|\, \nu_j\delta_{jk} \,\| \tag{5.5}$$

where the ν_j are the characteristic zeros of $\bar{\boldsymbol{P}}'\boldsymbol{\Gamma}\boldsymbol{P}$. Now let $\boldsymbol{Q} = \boldsymbol{P}\boldsymbol{U}$. Then (5.4) implies (5.1), and (5.5) implies

$$\bar{\boldsymbol{Q}}'\boldsymbol{\Gamma}\boldsymbol{Q} = \|\, \nu_j\delta_{jk} \,\| \quad . \tag{5.6}$$

Thus to establish (5.2) it remains but to prove that the ν_k, $1 \leqq k \leqq n$, are the roots of the characteristic equation $\det(\boldsymbol{\Gamma} - \nu\boldsymbol{G}) = 0$. But

$$\begin{aligned} \det(\bar{\boldsymbol{P}}'\boldsymbol{\Gamma}\boldsymbol{P} - \nu\boldsymbol{I}) &= \det(\bar{\boldsymbol{P}}'\boldsymbol{\Gamma}\boldsymbol{P} - \nu\bar{\boldsymbol{P}}'\boldsymbol{G}\boldsymbol{P}) \\ &= (\det \bar{\boldsymbol{P}}')[\det(\boldsymbol{\Gamma} - \nu\boldsymbol{G})](\det \boldsymbol{P}) \quad . \end{aligned}$$

Thus the ν_k, $1 \leqq k \leqq n$, are identical with the μ_j, $1 \leqq j \leqq n$, and the truth of (5.2) is apparent.

To prove (5.3) we note that $\boldsymbol{M} = \bar{\boldsymbol{Q}}'\boldsymbol{\Gamma}\boldsymbol{Q}$ is nonnegative definite and hence $\mu_j \geqq 0$. Also, since

$$\bar{\boldsymbol{Q}}'(\boldsymbol{G} - \boldsymbol{\Gamma})\,\boldsymbol{Q} = \boldsymbol{I} - \boldsymbol{M}$$

is nonnegative definite by hypothesis, $1 - \mu_j \geqq 0$. Hence $0 \leqq \mu_j \leqq 1$, $1 \leqq j \leqq n$. Now

$$1 = \det \boldsymbol{I} = \det \bar{\boldsymbol{Q}}'\boldsymbol{G}\boldsymbol{Q} = (\det \boldsymbol{G})\,|\det \boldsymbol{Q}\,|^2$$

and

$$m = \det \boldsymbol{M} = \det \bar{\boldsymbol{Q}}'\boldsymbol{\Gamma}\boldsymbol{Q} = (\det \boldsymbol{\Gamma})\,|\det \boldsymbol{Q}\,|^2$$

where

$$m = \prod_{j=1}^{n} \mu_j \quad .$$

But $0 \leqq m \leqq 1$, and $\det \boldsymbol{\Gamma} = m \det \boldsymbol{G}$. Thus (5.3) is established.

Theorem 5.1. Consider the model $\mathscr{M}_{Tp}[\boldsymbol{z}, \boldsymbol{\beta}, \boldsymbol{H}, \boldsymbol{\Sigma}]$. Let $\boldsymbol{b}$ be the Markov estimate of $\boldsymbol{\beta}$ and $\boldsymbol{b}^*$ the least squares estimate of $\boldsymbol{\beta}$. Then a necessary and sufficient condition that $\boldsymbol{b} = \boldsymbol{b}^*$ identically in the observation vector $\boldsymbol{z}$ is that $\boldsymbol{H} = \boldsymbol{V}\boldsymbol{C}$ where $\boldsymbol{C}$ is a nonsingular $p \times p$ matrix and $\boldsymbol{V}$ is a $T \times p$ matrix of characteristic vectors of $\boldsymbol{\Sigma}$.

Proof. *Sufficiency*. Let $\boldsymbol{H} = \boldsymbol{V}\boldsymbol{C}$ where $\boldsymbol{V}$ and $\boldsymbol{C}$ are as defined in the statement of the theorem. Then

$$\boldsymbol{\Sigma}\boldsymbol{V} = \boldsymbol{V}\boldsymbol{D}$$

where $\boldsymbol{D}$ is a $p \times p$ diagonal matrix of characteristic numbers of $\boldsymbol{\Sigma}$ (corresponding to the characteristic vectors which are the columns of $\boldsymbol{V}$.) Thus

$$\boldsymbol{\Sigma}(\boldsymbol{H}\boldsymbol{C}^{-1}) = (\boldsymbol{H}\boldsymbol{C}^{-1})\,\boldsymbol{D}$$

or

$$\bar{\boldsymbol{H}}'\boldsymbol{\Sigma}^{-1} = \bar{\boldsymbol{C}}'\boldsymbol{D}^{-1}\bar{\boldsymbol{C}}'^{-1}\bar{\boldsymbol{H}}' \quad . \tag{5.7}$$

Now the ME of $\boldsymbol{\beta}$ is

$$\boldsymbol{b} = (\bar{\boldsymbol{H}}'\boldsymbol{\Sigma}^{-1}\boldsymbol{H})^{-1}\bar{\boldsymbol{H}}'\boldsymbol{\Sigma}^{-1}\boldsymbol{z}$$

[see (4.7)]. If we replace $\bar{\boldsymbol{H}}'\boldsymbol{\Sigma}^{-1}$ in $\boldsymbol{b}$ by (5.7), there results

$$\begin{aligned}\boldsymbol{b} &= (\bar{\boldsymbol{C}}'\boldsymbol{D}^{-1}\bar{\boldsymbol{C}}'^{-1}\bar{\boldsymbol{H}}'\boldsymbol{H})^{-1}\bar{\boldsymbol{C}}'\boldsymbol{D}^{-1}\bar{\boldsymbol{C}}'^{-1}\bar{\boldsymbol{H}}'\boldsymbol{z} \\ &= (\bar{\boldsymbol{H}}'\boldsymbol{H})^{-1}\bar{\boldsymbol{H}}'\boldsymbol{z}\end{aligned}$$

—which is $\boldsymbol{b}^*$, the LSE of $\boldsymbol{\beta}$ [see (4.8)].

Necessity. Let $\boldsymbol{b} = \boldsymbol{b}^*$ identically in the observation vector $\boldsymbol{z}$. Then

$$(\bar{\boldsymbol{H}}'\boldsymbol{\Sigma}^{-1}\boldsymbol{H})^{-1}\bar{\boldsymbol{H}}'\boldsymbol{\Sigma}^{-1} = (\bar{\boldsymbol{H}}'\boldsymbol{H})^{-1}\bar{\boldsymbol{H}}'$$

or

$$(\bar{\boldsymbol{H}}'\boldsymbol{H})(\bar{\boldsymbol{H}}'\boldsymbol{\Sigma}^{-1}\boldsymbol{H})^{-1}\bar{\boldsymbol{H}}' = \bar{\boldsymbol{H}}'\boldsymbol{\Sigma} \quad . \tag{5.8}$$

Let $\boldsymbol{G} = \bar{\boldsymbol{H}}'\boldsymbol{\Sigma}^{-1}\boldsymbol{H}$ and $\boldsymbol{\Gamma} = \bar{\boldsymbol{H}}'\boldsymbol{H}$. From (4.10) [see also the comments following (3.2)] we infer that $\boldsymbol{G}$ and $\boldsymbol{\Gamma}$ are positive definite matrices. Thus, in the notation of Lemma 5.1,

$$\boldsymbol{\Gamma}\boldsymbol{G}^{-1} = (\bar{\boldsymbol{Q}}'^{-1}\boldsymbol{\Delta}\boldsymbol{Q}^{-1})(\boldsymbol{Q}\bar{\boldsymbol{Q}}') = \bar{\boldsymbol{Q}}'^{-1}\boldsymbol{\Delta}\bar{\boldsymbol{Q}}'$$

where $\boldsymbol{\Delta}$ is a $p \times p$ diagonal matrix and $\boldsymbol{Q}$ is nonsingular. The transpose of (5.8) may therefore be written as

$$\boldsymbol{\Sigma}\boldsymbol{H} = \boldsymbol{H}(\boldsymbol{Q}\boldsymbol{\Delta}\boldsymbol{Q}^{-1})$$

and multiplying on the right by $\boldsymbol{Q}$,

$$\boldsymbol{\Sigma}(\boldsymbol{H}\boldsymbol{Q}) = (\boldsymbol{H}\boldsymbol{Q})\boldsymbol{\Delta} \quad .$$

Now let $\boldsymbol{V} = \boldsymbol{H}\boldsymbol{Q}$. Then

$$\boldsymbol{\Sigma V} = \boldsymbol{V\Delta} \quad .$$

Since $\boldsymbol{\Delta}$ is diagonal, $\boldsymbol{V}$ is a $T \times p$ matrix of characteristic vectors of $\boldsymbol{\Sigma}$, and the diagonal elements of $\boldsymbol{\Delta}$ are characteristic numbers of $\boldsymbol{\Sigma}$. Since $\boldsymbol{Q}$ is non-singular, we may let $\boldsymbol{C} = \boldsymbol{Q}^{-1}$. Then

$$\boldsymbol{H} = \boldsymbol{VQ}^{-1} = \boldsymbol{VC},$$

as we desired to prove.

6. Efficiency of the Least Squares Estimate

Consider the model $\mathscr{M}_{Tp}[\boldsymbol{z}, \boldsymbol{\beta}, \boldsymbol{H}, \boldsymbol{\Sigma}]$ of Section 4. Let $\boldsymbol{b}^*$ be the LSE of $\boldsymbol{\beta}$. Then we define the *efficiency* of $\boldsymbol{b}^*$ as

$$\text{Eff}\, \boldsymbol{b}^* = \frac{\det \mathscr{E}(\boldsymbol{b} - \boldsymbol{\beta})(\bar{\boldsymbol{b}} - \bar{\boldsymbol{\beta}})'}{\det \mathscr{E}(\boldsymbol{b}^* - \boldsymbol{\beta})(\bar{\boldsymbol{b}}^* - \bar{\boldsymbol{\beta}})'}$$

where $\boldsymbol{b}$ is the ME of $\boldsymbol{\beta}$. From (4.10) and (4.9)

$$\text{Eff}\, \boldsymbol{b}^* = \frac{\det^2 \bar{\boldsymbol{H}}'\boldsymbol{H}}{(\det \bar{\boldsymbol{H}}'\boldsymbol{\Sigma H})(\det \bar{\boldsymbol{H}}'\boldsymbol{\Sigma}^{-1}H)} \quad . \tag{6.1}$$

We assert that $\text{Eff}\, \boldsymbol{b}^* \leqq 1$.

Theorem 6.1. In the model $\mathscr{M}_{Tp}[\boldsymbol{z}, \boldsymbol{\beta}, \boldsymbol{H}, \boldsymbol{\Sigma}]$ let $\boldsymbol{b}^*$ be the least squares estimate of $\boldsymbol{\beta}$. Then

$$\text{Eff}\, \boldsymbol{b}^* \leqq 1 \quad .$$

Proof. By (4.9), (4.10), and (4.11)

$$\mathscr{E}(\boldsymbol{b}^* - \boldsymbol{\beta})(\bar{\boldsymbol{b}}^* - \bar{\boldsymbol{\beta}})' - \mathscr{E}(\boldsymbol{b} - \boldsymbol{\beta})(\bar{\boldsymbol{b}} - \bar{\boldsymbol{\beta}})'$$

is nonnegative definite while $\mathscr{E}(\boldsymbol{b}^* - \boldsymbol{\beta})(\bar{\boldsymbol{b}}^* - \bar{\boldsymbol{\beta}})'$ and $\mathscr{E}(\boldsymbol{b} - \boldsymbol{\beta})(\bar{\boldsymbol{b}} - \bar{\boldsymbol{\beta}})'$ are both positive definite. Hence by (5.3) of Lemma 5.1,

$$\det \mathscr{E}(\boldsymbol{b}^* - \boldsymbol{\beta})(\bar{\boldsymbol{b}}^* - \bar{\boldsymbol{\beta}})' \geqq \det \mathscr{E}(\boldsymbol{b} - \boldsymbol{\beta})(\bar{\boldsymbol{b}} - \bar{\boldsymbol{\beta}})'$$

or $\text{Eff}\, \boldsymbol{b}^* \leqq 1$.

We may express $\text{Eff}\, \boldsymbol{b}^*$ in a more canonical form. Let $\boldsymbol{W}$ be a unitary matrix with the property that

$$\bar{\boldsymbol{W}}'\boldsymbol{\Sigma W} = \boldsymbol{L}$$

where $\boldsymbol{L}$ is the diagonal matrix $\|\lambda_j \delta_{jk}\|$ of the characteristic zeros of $\boldsymbol{\Sigma}$. Let $\boldsymbol{K} = \bar{\boldsymbol{W}}'\boldsymbol{H}$. Then $\bar{\boldsymbol{K}}'\boldsymbol{K} = \bar{\boldsymbol{H}}'\boldsymbol{H}$ and (6.1) may be written as

$$\text{Eff } \boldsymbol{b}^* = \frac{\det^2 \bar{\boldsymbol{K}}'\boldsymbol{K}}{(\det \bar{\boldsymbol{K}}'\boldsymbol{L}\boldsymbol{K})(\det \bar{\boldsymbol{K}}'\boldsymbol{L}^{-1}\boldsymbol{K})} \cdot$$

In particular, if $p = 1$, then $\boldsymbol{K}' = \boldsymbol{k}' = \{k_1, \cdots, k_T\}$ is a T-dimensional vector and

$$\text{Eff } b^* = \frac{|\boldsymbol{k}|^4}{(\bar{\boldsymbol{k}}'\boldsymbol{L}\boldsymbol{k})(\bar{\boldsymbol{k}}'\boldsymbol{L}^{-1}\boldsymbol{k})}$$

$$= \frac{\left(\sum_{t=1}^{T} |k_t|^2\right)^2}{\left(\sum_{t=1}^{T} |k_t|^2 \lambda_t\right)\left(\sum_{t=1}^{T} |k_t|^2 \lambda_t^{-1}\right)} \cdot$$

In this case (that is, $p = 1$), a lower bound for Eff b^* in terms of the characteristic zeros of $\boldsymbol{\Sigma}$ may be found. In fact, we have the following theorem:

Theorem 6.2. In the model $\mathscr{M}_{T1}[z, \beta, \boldsymbol{H}, \boldsymbol{\Sigma}]$ let b^* be the least squares estimate of β. Let λ_1 be the smallest and λ_T the largest characteristic zero of $\boldsymbol{\Sigma}$. Then

$$\text{Eff } b^* \geqq \frac{4\lambda_1\lambda_T}{(\lambda_1 + \lambda_T)^2} \cdot \tag{6.2}$$

Theorem 6.2 is an immediate consequence of *Kantorovich's inequality*—which we prove next.

Theorem 6.3 (Kantorovich's Inequality). Let $\alpha_t \geqq 0$, $1 \leqq t \leqq T$, and $\sum_{t=1}^{T} \alpha_t = 1$. Let $0 < \lambda_1 \leqq \lambda_2 \leqq \cdots \leqq \lambda_T < \infty$. Then

$$\left(\sum_{t=1}^{T} \alpha_t \lambda_t\right)\left(\sum_{t=1}^{T} \alpha_t \lambda_t^{-1}\right) \leqq \frac{(\lambda_1 + \lambda_T)^2}{4\lambda_1\lambda_T} \cdot \tag{6.3}$$

Proof. If $\lambda_1 = \lambda_T$, the theorem is trivial. Assume, therefore, that $\lambda_T > \lambda_1 > 0$. Since $(x - \lambda_1)(x - \lambda_T) \leqq 0$ on $[\lambda_1, \lambda_T]$,

$$\frac{1}{x} \leqq \frac{\lambda_1 + \lambda_T - x}{\lambda_1\lambda_T} \tag{6.4}$$

for all $x \in [\lambda_1, \lambda_T]$; and since $[(\lambda_1 + \lambda_T) - 2x]^2 \geqq 0$ for all x,

$$[(\lambda_1 + \lambda_T) - x]x \leqq \frac{(\lambda_1 + \lambda_T)^2}{4} \tag{6.5}$$

for all x.

Now by (6.4) with $x = \lambda_t \in [\lambda_1, \lambda_T]$,

$$\frac{1}{\lambda_t} \leqq \frac{\lambda_1 + \lambda_T - \lambda_t}{\lambda_1 \lambda_T} \quad .$$

Hence, multiplying by α_t and summing over t yields

$$\sum_{t=1}^{T} \alpha_t \lambda_t^{-1} \leqq \frac{1}{\lambda_1 \lambda_T}\left[(\lambda_1 + \lambda_T) - \sum_{t=1}^{T} \alpha_t \lambda_t\right] \quad . \tag{6.6}$$

From (6.5) with $x = \sum_{t=1}^{T} \alpha_t \lambda_t$

$$\left[(\lambda_1 + \lambda_T) - \sum_{t=1}^{T} \alpha_t \lambda_t\right]\left(\sum_{t=1}^{T} \alpha_t \lambda_t\right) \leqq \frac{(\lambda_1 + \lambda_T)^2}{4} \quad . \tag{6.7}$$

Equations (6.6) and (6.7) imply (6.3).

If we let

$$\alpha_t = \frac{|k_t|^2}{\sum_{s=1}^{T} |k_s|^2} \quad , \qquad 1 \leqq t \leqq T \quad ,$$

the truth of Theorem 6.2 becomes apparent.

Suppose now we consider the model $\mathscr{M}_{T1}[z, \beta, H, \sigma^2 I]$, that is, the case where the covariance matrix $\boldsymbol{\Sigma}$ is a multiple of the identity matrix. Then $\lambda_1 = \lambda_T$ and (6.2) becomes

$$\text{Eff}\, b^* = 1 \quad .$$

This, of course, corresponds to the case where the ME is identical with the LSE.

7. Maximum Likelihood Estimators

Consider the model $\mathscr{M}_{Tp}[\boldsymbol{z}, \boldsymbol{\beta}, \boldsymbol{H}, \sigma^2 \boldsymbol{I}]$ where σ^2 as well as $\boldsymbol{\beta}$ is to be considered as an unknown parameter. We shall assume that $\boldsymbol{z}$ is normal $(\boldsymbol{H\beta}, \sigma^2 \boldsymbol{I})$. Then the likelihood function of $\boldsymbol{z}$ is

$$l(\boldsymbol{z}) = \frac{1}{\pi^T \sigma^{2T}} \exp\left(\frac{-|\boldsymbol{z} - \boldsymbol{H\beta}|^2}{\sigma^2}\right) \quad . \tag{7.1}$$

[See (1.1) of Chapter III. We recall that if $\boldsymbol{R} = \sigma^2 \boldsymbol{I}$, then (1.9) of Chapter III implies that Re $\boldsymbol{z}$ and Im $\boldsymbol{z}$ are independent, and normal (Re $\boldsymbol{H\beta}$, $\frac{1}{2}\sigma^2 \boldsymbol{I}$) and (Im $\boldsymbol{H\beta}$, $\frac{1}{2}\sigma^2 \boldsymbol{I}$), respectively.] We wish to determine the *maximum likelihood estimate* (MLE) of $\boldsymbol{\beta}$ and σ^2.

Now

$$\frac{\partial}{\partial \boldsymbol{\beta}} \log l(\boldsymbol{z}) = -\frac{1}{\sigma^2}(-\boldsymbol{H}'\bar{\boldsymbol{z}} + \bar{\boldsymbol{A}}\bar{\boldsymbol{\beta}})$$

where $\boldsymbol{A} = \bar{\boldsymbol{H}}'\boldsymbol{H}$. Setting the above derivative equal to the p-dimensional zero vector leads to

$$\hat{\boldsymbol{\beta}} = \boldsymbol{A}^{-1}\bar{\boldsymbol{H}}'\boldsymbol{z}$$

as the MLE of $\boldsymbol{\beta}$. But [see (3.2)], this estimate is identical with the LSE $\boldsymbol{b}^*$ of $\boldsymbol{\beta}$. Thus we conclude that $\hat{\boldsymbol{\beta}} = \boldsymbol{b}^*$ when $\boldsymbol{z}$ is normally distributed with covariance matrix a multiple of $\boldsymbol{I}$.

To determine $\hat{\sigma}^2$, the MLE of σ^2, we differentiate the logarithm of the likelihood function with respect to σ^2:

$$\frac{\partial}{\partial \sigma^2} \log l(\boldsymbol{z}) = -\frac{T}{\sigma^2} + \frac{|\boldsymbol{z} - \boldsymbol{H}\boldsymbol{\beta}|^2}{\sigma^4} \quad .$$

Setting this derivative equal to zero we obtain

$$\hat{\sigma}^2 = \frac{1}{T}|\boldsymbol{z} - \boldsymbol{H}\hat{\boldsymbol{\beta}}|^2 \quad .$$

Thus we have shown:

Theorem 7.1. Consider the model $\mathscr{M}_{Tp}[\boldsymbol{z}, \boldsymbol{\beta}, \boldsymbol{H}, \sigma^2\boldsymbol{I}]$ where $\boldsymbol{\beta}$ and σ^2 are unknown parameters. Let $\boldsymbol{z}$ be normal $(\boldsymbol{H}\boldsymbol{\beta}, \sigma^2\boldsymbol{I})$. Then

$$\hat{\boldsymbol{\beta}} = \boldsymbol{A}^{-1}\bar{\boldsymbol{H}}'\boldsymbol{z} \tag{7.2}$$

and

$$\hat{\sigma}^2 = \frac{1}{T}|\boldsymbol{z} - \boldsymbol{H}\hat{\boldsymbol{\beta}}|^2 \tag{7.3}$$

are the maximum likelihood estimates of $\boldsymbol{\beta}$ and σ^2, respectively, where $\boldsymbol{A} = \bar{\boldsymbol{H}}'\boldsymbol{H}$. Furthermore,

$$\hat{\boldsymbol{\beta}} = \boldsymbol{b}^*$$

where $\boldsymbol{b}^*$ is the least squares estimate of $\boldsymbol{\beta}$.

The residual $\boldsymbol{z} - \boldsymbol{H}\boldsymbol{\beta}$ may be written

$$\boldsymbol{z} - \boldsymbol{H}\boldsymbol{\beta} = \boldsymbol{z} - \boldsymbol{H}\boldsymbol{b}^* + \boldsymbol{H}\boldsymbol{b}^* - \boldsymbol{H}\boldsymbol{\beta} \quad .$$

Thus the exponent in the likelihood function (7.1) becomes

$$\begin{aligned} |\boldsymbol{z} - \boldsymbol{H}\boldsymbol{\beta}|^2 = |\boldsymbol{z} - \boldsymbol{H}\boldsymbol{b}^*|^2 &+ (\bar{\boldsymbol{b}}^* - \bar{\boldsymbol{\beta}})'\boldsymbol{A}(\boldsymbol{b}^* - \boldsymbol{\beta}) \\ &+ (\bar{\boldsymbol{b}}^* - \bar{\boldsymbol{\beta}})'\bar{\boldsymbol{H}}'(\boldsymbol{z} - \boldsymbol{H}\boldsymbol{b}^*) + (\bar{\boldsymbol{z}} - \bar{\boldsymbol{H}}\bar{\boldsymbol{b}}^*)'\boldsymbol{H}(\boldsymbol{b}^* - \boldsymbol{\beta}) \quad . \end{aligned}$$

But

$$\bar{\boldsymbol{H}}'(\boldsymbol{z} - \boldsymbol{H}\boldsymbol{b}^*) = \bar{\boldsymbol{H}}'\boldsymbol{z} - A\boldsymbol{b}^* = \bar{\boldsymbol{H}}'\boldsymbol{z} - A(A^{-1}\bar{\boldsymbol{H}}'\boldsymbol{z}) = \boldsymbol{0} \quad .$$

Thus

$$\begin{aligned} |\boldsymbol{z} - \boldsymbol{H}\boldsymbol{\beta}|^2 &= |\boldsymbol{z} - \boldsymbol{H}\boldsymbol{b}^*|^2 + (\bar{\boldsymbol{b}}^* - \bar{\boldsymbol{\beta}})'A(\boldsymbol{b}^* - \boldsymbol{\beta}) \\ &= T\hat{\sigma}^2 + (\bar{\boldsymbol{b}}^* - \bar{\boldsymbol{\beta}})'A(\boldsymbol{b}^* - \boldsymbol{\beta}) \end{aligned} \tag{7.4}$$

by (7.3), and the likelihood function $l(\boldsymbol{z})$ may be written

$$l(\boldsymbol{z}) = (\pi\sigma^2)^{-T} \exp[-T\hat{\sigma}^2/\sigma^2] \exp[-\sigma^{-2}(\bar{\boldsymbol{b}}^* - \bar{\boldsymbol{\beta}})'A(\boldsymbol{b}^* - \boldsymbol{\beta})] \quad .$$

An examination of the formula above leads to the following conclusion.

Theorem 7.2. Consider the model $\mathscr{M}_{Tp}[\boldsymbol{z}, \boldsymbol{\beta}, \boldsymbol{H}, \sigma^2\boldsymbol{I}]$ where $\boldsymbol{\beta}$ and σ^2 are unknown parameters. Let $\boldsymbol{z}$ be normal$(\boldsymbol{H}\boldsymbol{\beta}, \sigma^2\boldsymbol{I})$. Then $\hat{\boldsymbol{\beta}}$ and $\hat{\sigma}^2$, the maximum likelihood estimates of $\boldsymbol{\beta}$ and σ^2 respectively, form a set of sufficient statistics for estimating $\boldsymbol{\beta}$ and σ^2.

Since $\mathscr{E}\boldsymbol{z} = \boldsymbol{H}\boldsymbol{\beta}$, equation (7.2) implies that

$$\mathscr{E}\hat{\boldsymbol{\beta}} = A^{-1}\bar{\boldsymbol{H}}'\mathscr{E}\boldsymbol{z} = A^{-1}A\boldsymbol{\beta} = \boldsymbol{\beta} \quad ,$$

that is, that $\hat{\boldsymbol{\beta}}$ is an unbiased estimate of $\boldsymbol{\beta}$. (This we already knew since $\boldsymbol{b}^* = \hat{\boldsymbol{\beta}}$ and the LSE is unbiased.) Now let us compute $\mathscr{E}\hat{\sigma}^2$. From (7.4)

$$T\mathscr{E}\hat{\sigma}^2 = \mathscr{E}|\boldsymbol{z} - \boldsymbol{H}\boldsymbol{\beta}|^2 - \mathscr{E}(\bar{\boldsymbol{b}}^* - \bar{\boldsymbol{\beta}})'A(\boldsymbol{b}^* - \boldsymbol{\beta}) \quad . \tag{7.5}$$

But

$$\mathscr{E}|\boldsymbol{z} - \boldsymbol{H}\boldsymbol{\beta}|^2 = \mathscr{E}\,\mathrm{tr}\,(\boldsymbol{z} - \boldsymbol{H}\boldsymbol{\beta})(\bar{\boldsymbol{z}} - \bar{\boldsymbol{H}}\bar{\boldsymbol{\beta}})' = \mathrm{tr}\,\sigma^2\boldsymbol{I} = T\sigma^2 \tag{7.6}$$

since $\boldsymbol{z}$ is normal $(\boldsymbol{H}\boldsymbol{\beta}, \sigma^2\boldsymbol{I})$ by hypothesis; and

$$\begin{aligned} \mathscr{E}(\bar{\boldsymbol{b}}^* - \bar{\boldsymbol{\beta}})'A(\boldsymbol{b}^* - \boldsymbol{\beta}) &= \mathscr{E}\,\mathrm{tr}\,(\boldsymbol{b}^* - \boldsymbol{\beta})(\bar{\boldsymbol{b}}^* - \bar{\boldsymbol{\beta}})'A \\ &= \mathscr{E}\,\mathrm{tr}\,\sigma^2\boldsymbol{A}^{-1}\boldsymbol{A} = p\sigma^2 \end{aligned} \tag{7.7}$$

since [see (4.12)] the covariance matrix of $\boldsymbol{b}^*$ is $\sigma^2\boldsymbol{A}^{-1}$. Using (7.6) and (7.7) in (7.5) gives us the desired result, namely,

$$\mathscr{E}\hat{\sigma}^2 = \frac{T - p}{T}\sigma^2 \quad .$$

Thus we have the following theorem.

Theorem 7.3. Consider the model $\mathscr{M}_{Tp}[\boldsymbol{z}, \boldsymbol{\beta}, \boldsymbol{H}, \sigma^2\boldsymbol{I}]$ where $\boldsymbol{\beta}$ and σ^2 are unknown parameters. Let $\boldsymbol{z}$ be normal $(\boldsymbol{H\beta}, \sigma^2\boldsymbol{I})$. Let $\hat{\boldsymbol{\beta}}$ and $\hat{\sigma}^2$ be the maximum likelihood estimates of $\boldsymbol{\beta}$ and σ^2, respectively. Then $\hat{\boldsymbol{\beta}}$ is an unbiased estimate of $\boldsymbol{\beta}$ and

$$s^2 = \frac{T}{T-p}\,\hat{\sigma}^2$$

is an unbiased estimate of σ^2.

Consider once again the model $\mathscr{M}_{Tp}[\boldsymbol{z}, \boldsymbol{\beta}, \boldsymbol{H}, \sigma^2\boldsymbol{I}]$ where $\boldsymbol{\beta}$ and σ^2 are unknown parameters. Let $\boldsymbol{z}$ be normal $(\boldsymbol{H\beta}, \sigma^2\boldsymbol{I})$. Let $\boldsymbol{b}^* = \hat{\boldsymbol{\beta}}$ and $s^2 = [T/(T-p)]\,\hat{\sigma}^2$ where $\hat{\boldsymbol{\beta}}$ and $\hat{\sigma}^2$ are the MLE of $\boldsymbol{\beta}$ and σ^2, respectively. We have seen in Theorem 7.3 that $\boldsymbol{b}^*$ and s^2 are unbiased. What we wish to do now is to show that $\boldsymbol{b}^*$ is normally distributed and that s^2 is chi-square distributed (see Theorem 7.4).

The first part is easy. Since

$$\boldsymbol{b}^* = \boldsymbol{A}^{-1}\bar{\boldsymbol{H}}'\boldsymbol{z}$$

is a linear combination of $\boldsymbol{z}$, and $\boldsymbol{z}$ is normal, it follows that $\boldsymbol{b}^*$ is normal with mean

$$\mathscr{E}\boldsymbol{b}^* = \boldsymbol{\beta}$$

and covariance

$$\mathscr{E}(\boldsymbol{b}^* - \boldsymbol{\beta})(\bar{\boldsymbol{b}}^* - \bar{\boldsymbol{\beta}})' = \sigma^2\boldsymbol{A}^{-1} \quad .$$

We shall now demonstrate that $[(T-p)/\frac{1}{2}\sigma^2]s^2$ is $\chi^2_{2(T-p)}$. Toward this end, choose a nonsingular $p \times p$ matrix $\boldsymbol{F}$ such that

$$\bar{\boldsymbol{F}}'\boldsymbol{A}\boldsymbol{F} = \boldsymbol{I}_p$$

where $\boldsymbol{I}_p$ is the $p \times p$ identity matrix. Since $\boldsymbol{A} = \bar{\boldsymbol{H}}'\boldsymbol{H}$, we may write the equation above as

$$(\bar{\boldsymbol{H}}\bar{\boldsymbol{F}})'(\boldsymbol{H}\boldsymbol{F}) = \boldsymbol{I}_p \quad .$$

Let $\boldsymbol{U}_1 = \boldsymbol{HF}$ and let $\boldsymbol{U}$ be a $T \times T$ unitary matrix whose first p columns are $\boldsymbol{U}_1$. We write

$$\boldsymbol{U} = \| \boldsymbol{U}_1 \quad \boldsymbol{U}_2 \|$$

where $\boldsymbol{U}_2$ is a $T \times (T-p)$ matrix. Because of the orthogonality of the columns of $\boldsymbol{U}$,

$$\begin{aligned} \bar{\boldsymbol{U}}_1'\boldsymbol{U}_1 &= \boldsymbol{I}_p \ , & \bar{\boldsymbol{U}}_1'\boldsymbol{U}_2 &= \boldsymbol{0} \ , \\ \bar{\boldsymbol{U}}_2'\boldsymbol{U}_1 &= \boldsymbol{0} \ , & \bar{\boldsymbol{U}}_2'\boldsymbol{U}_2 &= \boldsymbol{I}_{T-p} \ , \end{aligned} \tag{7.8}$$

where $\boldsymbol{I}_{T-p}$ is the $(T - p) \times (T - p)$ identity matrix.

Now let

$$\boldsymbol{v} = \bar{\boldsymbol{U}}'(\boldsymbol{z} - \boldsymbol{H}\boldsymbol{\beta}) \quad .$$

Then

$$\begin{aligned} \bar{\boldsymbol{v}}'\boldsymbol{v} &= (\bar{\boldsymbol{z}} - \bar{\boldsymbol{H}}\bar{\boldsymbol{\beta}})'\boldsymbol{U}\bar{\boldsymbol{U}}'(\boldsymbol{z} - \boldsymbol{H}\boldsymbol{\beta}) \\ &= |\boldsymbol{z} - \boldsymbol{H}\boldsymbol{\beta}|^2 \end{aligned} \tag{7.9}$$

and

$$\begin{aligned} (\bar{\boldsymbol{b}}^* - \bar{\boldsymbol{\beta}})'A(\boldsymbol{b}^* - \boldsymbol{\beta}) &= (\bar{A}^{-1}\boldsymbol{H}'\bar{\boldsymbol{z}} - \bar{A}^{-1}\boldsymbol{H}'\bar{\boldsymbol{H}}\bar{\boldsymbol{\beta}})' A(A^{-1}\bar{\boldsymbol{H}}'\boldsymbol{z} - A^{-1}\bar{\boldsymbol{H}}'\boldsymbol{H}\boldsymbol{\beta}) \\ &= (\bar{\boldsymbol{v}}'\bar{\boldsymbol{U}}')(\boldsymbol{U}_1\boldsymbol{F}^{-1})(\bar{\boldsymbol{F}}'^{-1}\bar{\boldsymbol{U}}_1'\boldsymbol{U}_1\boldsymbol{F}^{-1})^{-1}(\bar{\boldsymbol{F}}'^{-1}\bar{\boldsymbol{U}}_1')(\boldsymbol{U}\boldsymbol{v}) \end{aligned}$$

since $\boldsymbol{H} = \boldsymbol{U}_1\boldsymbol{F}^{-1}$ by definition of $\boldsymbol{U}_1$. Using (7.8), the expression above reduces to

$$(\bar{\boldsymbol{b}}^* - \bar{\boldsymbol{\beta}})'A(\boldsymbol{b}^* - \boldsymbol{\beta}) = \bar{\boldsymbol{v}}' \begin{bmatrix} \boldsymbol{I}_p & \boldsymbol{0} \\ \boldsymbol{0} & \boldsymbol{0} \end{bmatrix} \boldsymbol{v} \quad . \tag{7.10}$$

Let

$$\boldsymbol{v}^{*\prime} = \{v_{p+1}, \cdot \cdot \cdot \cdot, v_T\}$$

be the last $T - p$ components of $\boldsymbol{v}$. Then from (7.9) and (7.10)

$$|\boldsymbol{z} - \boldsymbol{H}\boldsymbol{\beta}|^2 - (\bar{\boldsymbol{b}}^* - \bar{\boldsymbol{\beta}})'A(\boldsymbol{b}^* - \boldsymbol{\beta}) = |\boldsymbol{v}^*|^2 \quad .$$

But from (7.4)

$$T\hat{\sigma}^2 = |\boldsymbol{v}^*|^2 \quad .$$

Since $\boldsymbol{z}$ is normal $(\boldsymbol{H}\boldsymbol{\beta}, \sigma^2\boldsymbol{I})$ and $\boldsymbol{v} = \bar{\boldsymbol{U}}'(\boldsymbol{z} - \boldsymbol{H}\boldsymbol{\beta})$ is a linear combination of $\boldsymbol{z}$, it follows that $\boldsymbol{v}$ is normal with mean

$$\mathscr{E}\boldsymbol{v} = \bar{\boldsymbol{U}}'\mathscr{E}(\boldsymbol{z} - \boldsymbol{H}\boldsymbol{\beta}) = \boldsymbol{0}$$

and covariance matrix

$$\mathscr{E}\boldsymbol{v}\bar{\boldsymbol{v}}' = \bar{\boldsymbol{U}}'\mathscr{E}(\boldsymbol{z} - \boldsymbol{H}\boldsymbol{\beta})(\bar{\boldsymbol{z}} - \bar{\boldsymbol{H}}\bar{\boldsymbol{\beta}})'\boldsymbol{U} = \sigma^2\bar{\boldsymbol{U}}'\boldsymbol{I}_T\boldsymbol{U} = \sigma^2\boldsymbol{I}_T \quad ,$$

where $\boldsymbol{I}_T$ is the $T \times T$ identity matrix. Thus $\boldsymbol{v}^*$ is normal $(\boldsymbol{0}, \sigma^2\boldsymbol{I}_{T-p})$ and $|\boldsymbol{v}^*|^2/\frac{1}{2}\sigma^2$ is $\chi^2_{2(T-p)}$ since $|\boldsymbol{v}^*|^2$ is the sum of the squares of $2(T - p)$ independent real Gaussian variates, each normal $(0, \frac{1}{2}\sigma^2)$. Hence

$$\frac{(T-p)s^2}{\frac{1}{2}\sigma^2} = \frac{T\hat{\sigma}^2}{\frac{1}{2}\sigma^2} = \frac{|\boldsymbol{v}^*|^2}{\frac{1}{2}\sigma^2}$$

is chi-square distributed with $2(T - p)$ degress of freedom.

Summarizing:

Theorem 7.4. Consider the model $\mathscr{M}_{Tp}[\boldsymbol{z}, \boldsymbol{\beta}, \boldsymbol{H}, \sigma^2\boldsymbol{I}]$ where $\boldsymbol{\beta}$ and σ^2 are unknown parameters. Let $\hat{\boldsymbol{\beta}}$ and $\hat{\sigma}^2$ be the maximum likelihood estimates of $\boldsymbol{\beta}$ and σ^2, respectively. Then $\hat{\boldsymbol{\beta}}$ is normally distributed $(\boldsymbol{\beta}, \sigma^2\boldsymbol{A}^{-1})$ where $\boldsymbol{A} = \bar{\boldsymbol{H}}'\boldsymbol{H}$, and $T\hat{\sigma}^2/\frac{1}{2}\sigma^2$ is chi-square distributed with $2(T - p)$ degrees of freedom. Furthermore, $\hat{\boldsymbol{\beta}}$ and $T\hat{\sigma}^2/\frac{1}{2}\sigma^2$ are independent.

We have proved, above, all but the last statement. Since $\boldsymbol{b}^* = \hat{\boldsymbol{\beta}}$ and $\boldsymbol{v}^*$ (where $|\boldsymbol{v}^*|^2 = T\hat{\sigma}^2$) are normal, we need only show that $\boldsymbol{b}^*$ and $\boldsymbol{v}^*$ are uncorrelated. Now

$$\begin{aligned} \mathscr{E}(\boldsymbol{b}^* - \boldsymbol{\beta})\,\bar{\boldsymbol{v}}^{*\prime} &= \mathscr{E}(\boldsymbol{A}^{-1}\bar{\boldsymbol{H}}'\boldsymbol{z} - \boldsymbol{A}^{-1}\bar{\boldsymbol{H}}'\boldsymbol{H}\boldsymbol{\beta})[\boldsymbol{U}_2'(\bar{\boldsymbol{z}} - \bar{\boldsymbol{H}}\bar{\boldsymbol{\beta}})]' \\ &= \boldsymbol{A}^{-1}\bar{\boldsymbol{H}}'\mathscr{E}(\boldsymbol{z} - \boldsymbol{H}\boldsymbol{\beta})(\bar{\boldsymbol{z}} - \bar{\boldsymbol{H}}\bar{\boldsymbol{\beta}})'\boldsymbol{U}_2 \\ &= \sigma^2\boldsymbol{A}^{-1}\bar{\boldsymbol{H}}'\boldsymbol{U}_2 \quad . \end{aligned}$$

But by definition of $\boldsymbol{U}_1$,

$$\bar{\boldsymbol{H}}' = \bar{\boldsymbol{F}}'^{-1}\bar{\boldsymbol{U}}_1' \quad .$$

Thus

$$\mathscr{E}(\boldsymbol{b}^* - \boldsymbol{\beta})\bar{\boldsymbol{v}}^{*\prime} = \sigma^2\boldsymbol{A}^{-1}\bar{\boldsymbol{F}}'^{-1}(\bar{\boldsymbol{U}}_1'\boldsymbol{U}_2) = \boldsymbol{0}$$

by (7.8). Hence $\boldsymbol{b}^*$ and s^2 are independently distributed.

8. Regression Curves

Let

$$\boldsymbol{z} = \begin{bmatrix} \boldsymbol{z}_p \\ \boldsymbol{z}_q \end{bmatrix} \tag{8.1}$$

be a partitioned random vector. Then the conditional expectation $\mathscr{E}(\boldsymbol{z}_p | \boldsymbol{z}_q)$ considered as a function of $\boldsymbol{z}_q$ is called the *regression function* of $\boldsymbol{z}_p$ on $\boldsymbol{z}_q$, and the equation

$$\boldsymbol{z}_p = \mathscr{E}(\boldsymbol{z}_p | \boldsymbol{z}_q)$$

is called the *regression curve* for $\boldsymbol{z}_p$. If $\mathscr{E}(\boldsymbol{z}_p | \boldsymbol{z}_q)$ is a linear function of $\boldsymbol{z}_q$, then the coefficients of the components of $\boldsymbol{z}_q$ in $\mathscr{E}(\boldsymbol{z}_p | \boldsymbol{z}_q)$ are called the *regression coefficients.*

If we choose a matrix $\boldsymbol{K}$ and a vector $\boldsymbol{a}$ such that the expectation

$$v = \mathscr{E}\,|(\boldsymbol{z}_p - \boldsymbol{c}_p) - [\boldsymbol{K}(\boldsymbol{z}_q - \boldsymbol{c}_q) + \boldsymbol{a}]|^2 \tag{8.2}$$

is minimized (where $\boldsymbol{c}_p = \mathscr{E}\boldsymbol{z}_p$, $\boldsymbol{c}_q = \mathscr{E}\boldsymbol{z}_q$) then we call

$$\boldsymbol{z}_p = \boldsymbol{K}(\boldsymbol{z}_q - \boldsymbol{c}_q) + \boldsymbol{c}_p + \boldsymbol{a} \tag{8.3}$$

the *mean square regression plane* of $\boldsymbol{z}_p$ on $\boldsymbol{z}_q$. We shall show that if $\boldsymbol{z}$ is Gaussian, then the mean square regression plane is identical with the regression curve.

Suppose, then, that $\boldsymbol{z}$ is a $(p + q)$-dimensional random vector normal $(\boldsymbol{c}, \boldsymbol{R})$ where $\boldsymbol{R}$ is positive definite. We shall partition $\boldsymbol{R}$ as

$$\boldsymbol{R} = \begin{bmatrix} \boldsymbol{R}_{pp} & \boldsymbol{R}_{pq} \\ \boldsymbol{R}_{qp} & \boldsymbol{R}_{qq} \end{bmatrix} \tag{8.4}$$

and partition $\boldsymbol{c}$ as

$$\boldsymbol{c} = \begin{bmatrix} \boldsymbol{c}_p \\ \boldsymbol{c}_q \end{bmatrix}$$

where $\boldsymbol{z}_p$ is normal $(\boldsymbol{c}_p, \boldsymbol{R}_{pp})$ [and $\boldsymbol{z}_q$ is normal $(\boldsymbol{c}_q, \boldsymbol{R}_{qq})$]. We shall also partition $\boldsymbol{R}^{-1}$ as

$$\boldsymbol{R}^{-1} = \begin{bmatrix} \boldsymbol{S}_{pp} & \boldsymbol{S}_{pq} \\ \boldsymbol{S}_{qp} & \boldsymbol{S}_{qq} \end{bmatrix} \tag{8.5}$$

where $\boldsymbol{S}_{pp}$ is $p \times p$. Then the conditional density function of $\boldsymbol{z}_p$ given $\boldsymbol{z}_q$ is normal

$$(\boldsymbol{c}_p - \boldsymbol{S}_{pp}^{-1}\boldsymbol{S}_{pq}(\boldsymbol{z}_q - \boldsymbol{c}_q), \boldsymbol{S}_{pp}^{-1})$$

and the conditional expectation is

$$\mathscr{E}(\boldsymbol{z}_p \,|\, \boldsymbol{z}_q) = \boldsymbol{c}_p - \boldsymbol{S}_{pp}^{-1}\boldsymbol{S}_{pq}(\boldsymbol{z}_q - \boldsymbol{c}_q) \quad .$$

Thus, in the Gaussian case, the regression curve for $\boldsymbol{z}_p$ on $\boldsymbol{z}_q$ is the hyperplane

$$\boldsymbol{z}_p = -\boldsymbol{S}_{pp}^{-1}\boldsymbol{S}_{pq}(\boldsymbol{z}_q - \boldsymbol{c}_q) + \boldsymbol{c}_p \quad . \tag{8.6}$$

Let us now derive the equation of the mean square regression plane. From Section 2 and (8.2)

$$\frac{\partial v}{\partial \boldsymbol{a}} = \bar{\boldsymbol{a}}$$

and

$$\frac{\partial v}{\partial \boldsymbol{K}} = \bar{\boldsymbol{K}}\bar{\boldsymbol{R}}_{qq} - \bar{\boldsymbol{R}}_{pq} .$$

Setting these derivatives equal to zero leads to

$$\boldsymbol{a} = \boldsymbol{0}$$

and

$$\boldsymbol{K} = \boldsymbol{R}_{pq}\boldsymbol{R}_{qq}^{-1} .$$

Thus from (8.3)

$$\boldsymbol{z}_p = \boldsymbol{R}_{pq}\boldsymbol{R}_{qq}^{-1}(\boldsymbol{z}_q - \boldsymbol{c}_q) + \boldsymbol{c}_p$$

is the equation of the mean square regression plane. But from the identity $\boldsymbol{R}^{-1}\boldsymbol{R} = \boldsymbol{I}$ [see (8.4) and (8.5)],

$$\boldsymbol{S}_{pp}\boldsymbol{R}_{pq} + \boldsymbol{S}_{pq}\boldsymbol{R}_{qq} = \boldsymbol{0} .$$

Hence

$$\boldsymbol{R}_{pq}\boldsymbol{R}_{qq}^{-1} = -\boldsymbol{S}_{pp}^{-1}\boldsymbol{S}_{pq}$$

and [see (8.6)] the mean square regression plane is identical with the regression curve (hyperplane).

In a somewhat different context, consider the regression model $\mathscr{M}_{Tp}[\boldsymbol{z}, \boldsymbol{\beta}, \boldsymbol{H}, \boldsymbol{\Sigma}]$ in the form

$$\boldsymbol{z} = \boldsymbol{H}\boldsymbol{\beta} + \boldsymbol{u}$$

where $\boldsymbol{u}$ is the residual, $\mathscr{E}\boldsymbol{u} = \boldsymbol{0}$, $\mathscr{E}\boldsymbol{u}\bar{\boldsymbol{u}}' = \boldsymbol{\Sigma}$. Suppose we wish to find a linear unbiased estimate $\tilde{\boldsymbol{\beta}}$ of $\boldsymbol{\beta}$ such that

$$\mathscr{E}\,|\tilde{\boldsymbol{\beta}} - \boldsymbol{\beta}|^2 \tag{8.7}$$

is minimized. We shall show that this approach yields an estimate $\tilde{\boldsymbol{\beta}}$ of $\boldsymbol{\beta}$ which is identical with the ME $\boldsymbol{b}$ of $\boldsymbol{\beta}$.

Toward this end let $\boldsymbol{K}$ be a $p \times T$ matrix. We wish to determine $\boldsymbol{K}$ such that

$$\tilde{\boldsymbol{\beta}} = \boldsymbol{K}\boldsymbol{z} \tag{8.8}$$

is a minimum variance [in the sense of (8.7)] unbiased estimate of $\boldsymbol{\beta}$. Now if $\tilde{\boldsymbol{\beta}}$ is to be unbiased we must have

$$\mathscr{E}\tilde{\boldsymbol{\beta}} = \boldsymbol{K}\mathscr{E}\boldsymbol{z} = \boldsymbol{K}\boldsymbol{H}\boldsymbol{\beta} = \boldsymbol{\beta}$$

or

$$\boldsymbol{KH} = \boldsymbol{I} \tag{8.9}$$

where $\boldsymbol{I}$ is the $p \times p$ identity matrix. Furthermore, the covariance matrix of $\tilde{\boldsymbol{\beta}}$ is

$$\mathscr{E}(\tilde{\boldsymbol{\beta}} - \boldsymbol{\beta})(\bar{\tilde{\boldsymbol{\beta}}} - \bar{\boldsymbol{\beta}})' = \boldsymbol{K\Sigma\bar{K}}'$$

and hence [see (8.7)]

$$\mathscr{E}|\tilde{\boldsymbol{\beta}} - \boldsymbol{\beta}|^2 = \operatorname{tr} \boldsymbol{K\Sigma\bar{K}}' \quad . \tag{8.10}$$

Now for any p-dimensional vector $\boldsymbol{\lambda}$,

$$\bar{\boldsymbol{\lambda}}'\boldsymbol{KH\lambda} = |\boldsymbol{\lambda}|^2$$

[see (8.9)] is a real scalar. Since the "variance" of $\tilde{\boldsymbol{\beta}}$ is $\operatorname{tr} \boldsymbol{K\Sigma\bar{K}}'$ [see (8.10)], we must minimize

$$v = \operatorname{tr} \boldsymbol{K\Sigma\bar{K}}' - \bar{\boldsymbol{\lambda}}'\boldsymbol{KH\lambda}$$

where $\boldsymbol{\lambda}$ plays the role of a Lagrange multiplier (see Section 4). Setting the derivative

$$\frac{dv}{d\boldsymbol{K}} = (\boldsymbol{\Sigma\bar{K}}')' - \bar{\boldsymbol{\lambda}}\boldsymbol{\lambda}'\boldsymbol{H}'$$

equal to the $p \times T$ zero matrix leads to

$$\boldsymbol{K} = \boldsymbol{\lambda}\bar{\boldsymbol{\lambda}}'\bar{\boldsymbol{H}}'\boldsymbol{\Sigma}^{-1} \quad . \tag{8.11}$$

Thus by (8.9)

$$\boldsymbol{I} = \boldsymbol{\lambda}\bar{\boldsymbol{\lambda}}'\bar{\boldsymbol{H}}'\boldsymbol{\Sigma}^{-1}\boldsymbol{H}$$

and

$$\boldsymbol{\lambda}\bar{\boldsymbol{\lambda}}' = (\bar{\boldsymbol{H}}'\boldsymbol{\Sigma}^{-1}\boldsymbol{H})^{-1} \tag{8.12}$$

(since $\bar{\boldsymbol{H}}'\boldsymbol{\Sigma}^{-1}\boldsymbol{H}$ is nonsingular.)

Thus from (8.8), (8.11), and (8.12)

$$\tilde{\boldsymbol{\beta}} = (\bar{\boldsymbol{H}}'\boldsymbol{\Sigma}^{-1}\boldsymbol{H})^{-1}\bar{\boldsymbol{H}}'\boldsymbol{\Sigma}^{-1}\boldsymbol{z}$$

—which is identical with the ME $\boldsymbol{b}$ of $\boldsymbol{\beta}$ [see (4.7)]. The "variance" of $\tilde{\boldsymbol{\beta}}$ is [see (8.10)]

$$\begin{aligned}\mathscr{E}|\tilde{\boldsymbol{\beta}} - \boldsymbol{\beta}|^2 &= \operatorname{tr} \boldsymbol{K\Sigma\bar{K}}' \\ &= \operatorname{tr} (\bar{\boldsymbol{H}}'\boldsymbol{\Sigma}^{-1}\boldsymbol{H})^{-1}\end{aligned}$$

[see (4.10)].

REFERENCES

[AS] M. Abramowitz and I. A. Stegun, *Handbook of Mathematical Functions with Formulas, Graphs, and Mathematical Tables,* AMS 55, Nat. Bur. Standards, U.S. Govt. Print. Office, 1964.

[Ai] A. C. Aitken, *Determinants and Matrices,* Interscience, New York, 1942.

[An] T. W. Anderson, *The Statistical Analysis of Time Series,* John Wiley, New York, 1971.

[Ar] R. Arens, *Complex processes for envelopes of normal noise,* Trans. I.R.E., **IT-3**, 204–207 (1957).

[BT] R. B. Blackman and J. W. Tukey, *The Measurement of Power Spectra,* Dover, New York, 1958.

[BG] H. G. Booker and W. E. Gordon, *A theory of radio scattering in the troposphere,* Pro. I.R.E., **38**, 401–412 (1950).

[BRS] H. G. Booker, J. A. Ratcliffe, and D. H. Shinn, *Diffraction from an irregular screen with applications to ionospheric problems,* Phil. Trans. Roy. Soc. London Ser. A, **242**, 579–609 (1950).

[CoL] E. A. Coddington and N. Levinson, *Theory of Ordinary Differential Equations,* McGraw-Hill, New York, 1955.

[Cr] H. Cramér, *Mathematical Methods of Statistics,* Princeton University Press, Princeton, 1946.

[CL] H. Cramér and M. R. Leadbetter, *Stationary and Related Stochastic Processes,* John Wiley, New York, 1967.

[D] J. Dugundji, *Envelopes and pre-envelopes of real waveforms,* Trans. I.R.E, **IT-4**, 53–57 (1958).

[EE] L. Ehrman and R. Esposito, *On the accuracy of the envelope method for the measurement of Doppler spread,* Trans. I.E.E.E., **COM-17,** 578–581 (1969).

[Go] N. R. Goodman, *Statistical analysis based on a certain multivariate complex Gaussian distribution (An introduction),* Ann. Math. Statist., **34**, 152–177 (1963).

[GR] U. Grenander and M. Rosenblatt, *Statistical Analysis of Stationary Time Series,* John Wiley, New York, 1957.

[Gr] T. L. Grettenberg, *A representation theorem for complex normal pro-*

cesses, Trans. I.E.E.E., **IT-11,** 305–306 (1965).

[H] E. J. Hannan, *Time Series Analysis,* Methuen, London, 1960.

[HC] R. V. Hogg and A. T. Craig, *Introduction to Mathematical Statistics,* Macmillan, New York, 1959.

[K] D. G. Kabe, *Complex analogues of some classical noncentral multivariate distributions,* Austral, J. Statist., **8**, 99–103 (1966).

[KRR 1] E. J. Kelly, I.S. Reed, and W. L. Root, *The detection of radar echoes in noise I,* J. Soc. Indust. Appl. Math., **8**, 309–341 (1960).

[KRR 2] E. J. Kelly, I. S. Reed, and W. L. Root, *The detection of radar echoes in noise II,* J. Soc. Indust. Appl. Math., **8**, 481–507 (1960).

[La] F. Lane, *Frequency effects in the radar return from turbulent, weakly ionized missile wakes,* A.I.A.A. J., **5**, 2193–2200 (1967).

[Le] M. J. Levin, *Power spectrum parameter estimation,* Trans. I.E.E.E., **IT-11**, 100–107 (1965).

[McG] W. F. McGee, *Complex Gaussian noise moments,* Trans. I.E.E.E., **IT-17**, 149–157 (1971).

[Mid] D. Middleton, *An Introduction to Statistical Communication Theory,* McGraw-Hill, New York, 1960.

[Mil 1] K. S. Miller, *Multidimensional Gaussian Distributions,* John Wiley, New York, 1964.

[Mil 2] K. S. Miller, *Complex Gaussian processes,* SIAM Rev., **11**, 544–567 (1969).

[Mil 3] K. S. Miller, *Some multivariate density functions of products of Gaussian variates,* Biometrika, **52**, 645–646 (1965).

[Mil 4] K. S. Miller, *Some multivariate t-distributions,* Ann. Math. Statist., **39**, 1605–1609 (1968).

[Mil 5] K. S. Miller, *Linear Difference Equations,* W. A. Benjamin, New York, 1968.

[Mil 6] K. S. Miller, *Nonstationary autoregressive processes,* Trans. I.E.E.E., **IT-15**, 315–316 (1969).

[Mil 7] K. S. Miller, *A note on stochastic difference equations,* Ann. Math. Statist., **39**, 270–271 (1968).

[Mil 8] K. S. Miller, *Linear Differential Equations in the Real Domain,* W. W. Norton, New York, 1963.

[Mil 9] K. S. Miller, *Properties of impulsive responses and Green's functions,* Trans. I.R.E., **CT-2**, 26–31 (1955).

[MR 1] K. S. Miller and M. M. Rochwarger, *On estimating spectral moments in the presence of colored noise,* Trans. I.E.E.E., **IT-16**, 303–309 (1970).

[MR 2] K. S. Miller and M. M. Rochwarger, *A covariance approach to spectral moment estimation,* Trans. I.E.E.E., **IT-18**, 588–596 (1972).

[MR 3] K. S. Miller and M. M. Rochwarger, *Estimation of spectral moments of time series,* Biometrika, **57**, 513–517 (1970).

[MS 1] K. S. Miller and H. Sackrowitz, *Relationships between biased and unbiased Rayleigh distributions,* SIAM J. Appl. Math., **15**, 1490–1495 (1967).

[MS 2] K. S. Miller and H. Sackrowitz, *Distributions associated with the quadrivariate normal,* Indust. Math., **15**, 1–14 (1965).

[Mu] M. E. Munroe, *Introduction to Measure and Integration,* Addison-Wesley, Reading, 1953.

[N] J. L. Neuringer, *Derivation of the spectral density function of the energy scattered from an underdense turbulent wake,* A.I.A.A. J., **7**, 728–730 (1969).

[Ra] J. A. Ratcliffe, *Diffraction from the ionosphere and the fading of radio waves,* Nature, **162**, 9–11 (1948).

[Ri] S. O. Rice, *Statistical fluctuations of radio field strength far beyond the horizon,* Proc. I.R.E., **41**, 274–281 (1953).

[S] D. J. Sakrison, *Efficient recursive estimation of the parameters of a radar or radio astronomy target,* Trans. I.E.E.E., **IT-12**, 35–41 (1966).

[SBS] M. Schwartz, W. R. Bennett, and S. Stein, *Communication Systems and Techniques,* McGraw-Hill, New York, 1966.

[Wi] D. V. Widder, *The Laplace Transform,* Princeton University Press, Princeton, 1941.

[Wo] R. A. Wooding, *The multivariate distribution of complex normal variables,* Biometrika, **43**, 212–215 (1956).

INDEX